D0984666

THE MIND AND
THE EYE

THE MIND AND THE EYE

A STUDY OF THE BIOLOGIST'S STANDPOINT

BY

AGNES ARBER

M.A., D.Sc., F.R.S.
F.L.S.

CAMBRIDGE
AT THE UNIVERSITY PRESS
1954

PUBLISHED BY
THE SYNDICS OF THE CAMBRIDGE UNIVERSITY PRESS

London Office: Bentley House, N.W.1
American Branch: New York

Agents for Canada, India, and Pakistan: Macmillan

Printed in Great Britain at the University Press, Cambridge
(Brooke Crutchley, University Printer)

To

D.S.R. and P.C.R.

sai quel che si tace

PREFACE

IN the course of a period, extending over half a century, in which my concern has been with research in plant morphology, I have found my mind dwelling more and more upon the nature of scientific thought, and its relation to other intellectual activities. Such ponderings have led me gradually to realize how little I, as a biologist, could actually justify, or even, indeed, understand, the nature of the basic assumptions and modes of argument which, in accordance with scientific tradition, I was taking simply as 'given'. For the last twenty years I have been attempting to clarify my ideas on these subjects, with the aid of such reading in metaphysics as is within the compass of the amateur. In *The Natural Philosophy of Plant Form*, published by the Cambridge University Press in 1950, I have touched upon certain aspects of the botanist's attitude to his work. The present book offers a more generalized analysis of the biologist's approach to his own subject and to philosophy. Of the defects and limitations of this study I am profoundly conscious; but my hope is that its very inadequacies may stimulate others to cast an illumination, more powerful than my rushlight, upon the biologist's road to reality.

AGNES ARBER

CAMBRIDGE
22 *May* 1953

ACKNOWLEDGEMENTS

I WISH to express my gratitude to those authors and publishers who have allowed me to use quotations from the following books: J. Clark and A. Geikie (1910), *Physical Science in the Time of Nero* (Macmillan and Co. Ltd.); F. M. Cornford (1935), *Plato's Theory of Knowledge*, and (1937) *Plato's Cosmology* (Routledge and Kegan Paul, Ltd.); E. R. Dodds (1923), *Select Passages Illustrating Neoplatonism* (S.P.C.K.); Pfeiffer's *Meister Eckhart* (1949), translated by C. de B. Evans (John M. Watkins); F. Galton (1889), *Natural Inheritance* (Macmillan and Co. Ltd.); B. Jowett (1871), *The Dialogues of Plato* (Clarendon Press, Oxford); C. Singer (1917), 'The scientific views and visions of Saint Hildegard', in *Studies in the History of Science*, vol. 1 (Clarendon Press, Oxford); D. W. Singer (1950), *Giordano Bruno* (Henry Schuman, Inc., New York); N. Kemp Smith (1923), *A Commentary to Kant's 'Critique of Pure Reason'*, and (1933), *Immanuel Kant's Critique of Pure Reason* (Macmillan and Co. Ltd.); A. Wolf (1910), *Spinoza's Short Treatise* (A. and C. Black, Ltd.).

I am also indebted to Henry Schuman, Inc., and to Professor Ashley Montagu and Dr George Sarton, for permission to incorporate in Chapter IV the substance of an article which I contributed to *Studies and Essays offered to George Sarton* (1946).

It is now forty years since the Cambridge University Press published my first book, and I should like to take this opportunity of offering my tribute to the Syndics and the Staff for the inexhaustible kindness and skill with which they smooth their authors' way.

A. A.

CONTENTS

CONTENTS

PART I
THE NATURE OF BIOLOGICAL RESEARCH

PART II
THE BASIS OF BIOLOGICAL THINKING

PART I

THE NATURE OF
BIOLOGICAL RESEARCH

INTRODUCTION

A BIOLOGIST, asked to formulate the aim of his studies in the broadest and most general terms, might answer that he wanted to know, and above all *to understand*, the form, structure and behaviour, of living things; the chemical and physical factors in their functioning; their development, reproduction, and genetics; their relation to the non-living environment and to one another; and their race history. He would, of course, make the reservation that his acquaintance with most of the labyrinthine detail of this enormous field could not be more than second-hand, derivative, and sketchy; his first-hand knowledge could relate only to a strictly limited region. For the purpose of offering some contribution of his own to science, still further limitation would be demanded, and he would have to choose, out of the area on which his interest mainly concentrated, some place where he felt there was a certain obscurity—concerning facts, or their interpretation, or both—which he could hope to enlighten. He might be fortunate enough to turn his attention to some point where either a clear-cut question could at once be put to Nature, or where, if he could achieve a more exact description of the facts, some individual inquiry might eventually emerge and formulate itself. This brings into view the research worker's first objective—the choice either of a specific question, or of some problematic area of biological thought from which, under the stimulus of factual study, he can hope that a distinct question will gradually crystallize. The finding of a genuine problem is indeed of fundamental importance. The mind languishes when faced with an enigma which lacks content, but, confronted by a problem worthy of its steel, it often displays powers which were previously quite unsuspected.[1]

When the researcher has arrived at a definite question, he

---

[1] Cf. Meyerson, É. (1931), vol. II, pp. 615–17.

I-2

hardly ever finds the facts, which he needs for the solution of his problem, ready to his hand; in order that they may be facts *for him*, he has to acquire them for himself, from his own standpoint. The second stage is thus the observational or experimental search for relevant data, and an attempt to organize them. After this comes the third stage—the endeavour to interpret the marshalled facts. The data may prove to fit neatly into the existing framework of scientific explanation, or they may demand a reconstruction of this framework before they can find their place in it.

If the programme has been carried successfully through these three stages, some sort of solution of the initial problem will have been reached, and the biologist comes to the fourth phase of his work—the attempt to test the validity of his solution; in other words he has to do what is commonly described as 'proving' its truth.

The first four phases just enumerated have had only an internal reference, but, in the next, the researcher has to turn his gaze outwards; for he has now to consider how to communicate the course of his investigations, and the conclusions to which he has come, to his fellow-students, who cannot know what it is all about, except through the medium of his skill with pen, and sometimes with pencil.

In these days of specialization, the biologist who hopes to add anything to science is bound to confine his attention during the greater part of his working life to these five stages. Nevertheless, most workers occasionally, and those of a contemplative temperament more frequently, are conscious of an urge to pass on to a further stage—the sixth in the sequence we are tracing. In this the biologist stands back from the individual jobs to which he has set his hand, in order to see them in the context of thought in general; to criticize their presuppositions and the mode of thinking which they employ; and to discover how the intellectual and sensory elements, which they include, are interconnected. This urge is more likely to supervene late in life, since the contemplative spirit then gathers a force that it could not collect in earlier years, when the mind, in its fresh receptivity, concen-

4

trated its attention upon the delightful detail of the factual multiplicity of living things. This terminal stage of biological thinking may take more than one form. The worker may attempt to realize his own individual findings in relation to those far-reaching problems which are common to the various fields of thought; his activities at this level will then extend into the territory of philosophy. On the other hand, if it is the visual rather than the purely intellectual aspect of his problems which enthrals him, and affords him a glimpse of wider horizons, he may discover that his reaction to his experiences is, in the long run, that of the artist rather than the philosopher. Vesalius is a cogent example of a great biologist, who did not think in terms of intellectual concepts, but was essentially a creative artist-naturalist.[1]

The pattern of the present book is so disposed as to follow, in general outline, the ascending scale suggested by the six stages here recognized. The chief stress will be laid upon the ultimate grade, in which, in the writer's personal view, the previous phases all find their end and their justification.

[1] On the position of Vesalius in biology, see Singer, C. (1925); Nordenskiöld, N. E. (1950); and Singer, C., and Rabin, C. (1946).

5

CHAPTER I

THE BIOLOGIST AND HIS PROBLEM

SINCE the first step in biological research involves the decision as to the question on which to concentrate, the researcher is at once put upon his mettle, for the full recognition and appreciation of a problem may task him even more severely than its solution. It is undoubtedly true that the "difficulty in most scientific work lies in framing the questions rather than in finding the answers";[1] an unerring instinct for the valid problem is likely, indeed, to be the ripened harvest of the scientific life, rather than its first-fruits. In practice, however, the paradox, that the thinker needs to have reached the end before he can make an effective beginning, must be ignored; for full preparedness for the start never comes until the time for starting has long gone by. So the biologist must begin without tarrying for full equipment, making shift from the first to pose his own question as best he may. It is true that the difficulty may be side-tracked temporarily, since the young worker, during his apprenticeship, is often supplied with a ready-made problem by his supervisor. Such tutelage is inevitable, if the student is to be put in possession of the traditional mode of approach and the existing technique. As a strictly ephemeral phase, it should not undermine originality—it may, on the contrary, assist it, for no man can create a new offshoot from tradition, or break with it to advantage, unless he knows precisely *what it is* that he is developing or discarding. The fact that spoon-feeding with imposed problems sometimes lingers on into later years, owing to the exigencies of team-work, may seriously impede intrinsic development; for each biologist ought to be able to say to himself, like Descartes, that his intention is to build upon a foundation that is all his own.[2] It was Descartes, again, with his sturdy

---

[1] Boycott, A. E. (1929), p. 95.
[2] "de bastir dans un fons qui est tout a moy" [Descartes, R.] (1637), p. 16; (1947), p. 15. Translation, Haldane, E. S., and Ross, G. R. T. (1911–12), vol. 1, p. 90.

individualism, who held that the only help that can be given to a researcher is to defray his needful scientific expenses, and then to ensure that no one filches his leisure.[1]

When we consider problems in the light of history, we find that they are, in one aspect, suprapersonal; for what particular biological questions excite special interest at any given period is a sign of the general intellectual focus of that period; contemporary topics and those of the past are thus seen through quite different spectacles. One becomes acutely conscious of this divergence in looking over old volumes of research journals; they are filled, in the main, with articles in which the authors tackle, with enthusiasm, problems which have now simply lost their interest for us. It is not by any means always that these problems have since found their answers, and that they have been built into the structure of the science; often they have been shelved rather than solved, yet the notion of pursuing them evokes in us now nothing but ennui. This feeling may have a sound basis, for a problem put aside in one period must have the right interval of dormancy before awakening, freshened, to an unforced solution, when the time is naturally ripe. In the history, for instance, of such a subject as psychology, which is relatively modern as a self-conscious and autonomous science, we can see how many starts have been made in different directions, and have afterwards been set aside in favour of other lines.[2] It seems safe to prophesy that future generations will, from a more advanced standpoint, return to reap a harvest from some of these now forsaken beginnings. That there is a time for everything, and a season for every purpose under heaven, is indeed continually exemplified in the history of science; the general intellectual atmosphere of any given moment has an effect upon this history which is compulsive to a humiliating degree. In every period certain classes of beliefs and ideas have been actively distasteful, and even workers of some independence of mind are found to have shrunk

---

[1] " ...fournir des frais des experiences dont il auroit besoin, et du reste empescher que son loisir ne luy fust osté par l'importunité de personne" [Descartes, R.] (1637), p. 73; (1947), p. 73. Translation, Haldane, E. S., and Ross, G. R. T. (1911–12), vol. I, p. 127.

[2] Cf. the history of a century of psychological study in Flugel, J. C. (1933).

from them as if they were tabooed. In the literature of biology one may occasionally detect a hint of an untrodden and un-authorized way, which held out a prospect of a fresh view-point; but the man, who had glimpsed it, too often proceeded to turn his back upon it, reverting to the familiar beaten paths, where he could absorb confidence from the reassuring society of his fellow-workers. The tyranny of the *Zeitgeist* is obvious enough to us in the scientific writing of fifty years ago, whereas we are less aware of it in that of this year, because we are always too much be-dazzled by contemporaneity to judge the present fairly; half a century hence, biologists will, no doubt, find as much tedium in the general approach of to-day as we do in that of a time which is to us equally long past. It is sad that it should be so, for we lose a great deal if we discard the conceptions of past generations indiscriminately, because the swing of the pendulum of fashion makes us react against them.

In every field of biological research, the intellectual climate and the general beliefs of the time cannot but be especially and rightfully influential where the beginner is concerned. He will best 'find himself' in his work if he embarks upon a task which he can realize as a part, even if a minimal part, of some attempt much larger than itself, which is a tributary of the main thought-stream of the period. If, for instance, he is a taxonomist, aiming only at dealing with the affinities of a single form, he will make the best job of his attempt by viewing it as a microcosm of the comprehensive subject of relationship in the whole living world, as seen on the background of existing classificatory ideas. Later on, he may profoundly modify, or discard these ideas, but mean-while the natural thing will be to begin by putting them to the test of use, so that, when he departs from them, it will be for well-knit reasons.

The work of the taxonomist, like that of the morphologist, is sometimes slighted as being purely 'descriptive', and hence of no theoretic interest; but this criticism is based on an inadequate notion of what the term *description* should mean in biology. By rights it should come into quite a different category from mere mechanical representation in words. To begin with, *what* is to be described is not fortuitous, but demands preliminary selection, involving theoretical implications; otherwise the observer would be lost in a chaotic nightmare of phenomena clamouring on all sides for his attention. Moreover, every description exists on a background of biological theory, to which it is intimately related—whether this relationship is expressed or merely understood. In so far as description includes this element of relatedness, it may be held to be contributing its mite towards testing the assumption that there is coherence between the different parts within Nature.[1] Work that is seemingly a descriptive study of the most pedestrian type may thus, on closer scrutiny, be seen to have also some theoretic function. At a higher level, the ideal of scientific research appears to be most nearly approached when delicate and detailed practical investigation is employed in a conscious attempt to solve theoretic questions of the broadest kind. Conspicuous examples of such work are seen in the still young study of what is called 'fine structure'—the realm of organization between individual molecules and microscopic structures[2]—and in the related 'chemical geography' within the cell.[3] Here we have a factual analysis of perceptual data obtained by a highly evolved technique, associated with conceptual synthesis of an abstract type. The study of form is thus raised to a plane above, but including, pure empiricism.

One of the factors determining the choice of a researcher's problem is whether the trend of his mind is towards the descriptive, or the more definitely theoretical. In science, almost as much as in art, tastes have to be reckoned with; for the researcher's choice of a field cannot be satisfying unless both feeling and intellect converge upon it. No one, indeed, can reach a creative solution of a problem which he does not approach *con amore*. Scientific men, of a non-analytic turn, when catechized as to *why* they do their work, often give elaborately altruistic reasons for their devoted concentration; but those who are more introspective confess that, in reality, they do it simply because they like it, and (in the Cornish sense) it 'belongs' to them to do

---

[1] See Chapter VII, pp. 82–8, of the present book.
[2] Cf., for example, Picken, L. E. R. (1950).
[3] Bradfield, J. R. G. (1950), p. 81.

it; they recognize that Kant was right when he pointed out that the pleasurableness of advancing knowledge easily gives it the appearance of being a duty.[1]

Another question of a highly personal kind, confronting the researcher, concerns the *relative size* of problem which can fitly be taken in hand. The idea that scale is of special significance in biological mechanics, has been fully realized only in recent times; a corresponding principle plays its part in mental life. No biologist can do his best work unless the scale of his problem is duly proportioned to the height of his powers and the width of his vision. For a man of considerable mental calibre, a problem with strictly limited range may prove hopelessly inhibiting, through giving a frustrating sense of lack of scope: Hercules was probably rather ineffectual with the distaff, during his servitude to Omphale.[2] By the ablest workers of all, however, this drawback is sometimes overcome, since, under their hands, what seemed to be a trivial question, may reveal unsuspected depths of significance. Originality will out; the Arab writers, who held themselves to be mere commentators upon Aristotle, were in fact producing a philosophy rich in new elements.[3] The many who are less gifted, may find, however, as Locke recognized long ago, that "'tis Ambition enough to be employed as an Under-Labourer in clearing the Ground a little, and removing some of the Rubbish, that lies in the way to Knowledge".[4] When, however, a researcher, who might have been happily and usefully employed on a minor undertaking of this kind, tries to struggle with a major question which dwarfs him, such powers as he can claim are paralysed, and he is prevented from doing the service to science that in him lies; or else his reaction takes the form of whittling and paring down the subject, until he can bring it within his competence by rejecting its essential difficulties. He then congratulates himself on the facile conquest of his problem,

---

[1] "Die Annehmlichkeit, welche die Erweiterung des Wissens begleitet, wird sehr leicht den Schein der Pflichtmässigkeit annehmen." From *Träume eines Geistersehers* (1766), in Kant, I. (1902–38), vol. II (1905), p. 369. For a study of this essay see Caird, E. (1889), vol. I, pp. 146 *et seq.*

[2] Cf. Coleridge, S. T. (1817), vol. II, p. 152.

[3] Cf., for instance, Renan, E. (1852), pp. 66–7.

[4] Locke, J. (1690), Epistle to the Reader.

forgetting that he induced it to disarm before he tried to overcome it. Indeed, the best of which a man is capable can be achieved only when his powers, and the theme on which he exercises them, are fitted to one another without discrepancy. This happy consummation may be attained only after a long and exacting process of trial and error. The difficulties and failures of the scientist in his search for the right topic, can be paralleled in other spheres. Coleridge (as a prose writer) and Samuel Johnson come into one's mind as men, who, despite their gigantic capacity, were helpless when it came to constructing an adequate and integrated framework for their own abilities; their mental activity thus never reaped its due harvest. One can recall instances of others who, though they eventually surmounted this initial difficulty, did so only after deliberate and prolonged search for a theme. In the three years before Gibbon's choice of *The Decline and Fall* was determined, he considered and rejected ten other historical subjects,[1] while Milton brooded over nearly one hundred alternatives before he finally decided upon *Paradise Lost*,[2] which gave him the very field his genius needed.

Problem choice is only a minor part of a wider intellectual discipline—the art of rejection. Confronted with a limitless field for possible inquiry, the researcher is bound to neglect, for his special purposes, all but a selected finite fragment. A choice is always a sacrifice;[3] but it is true, though paradoxical, that there is much that is positive in the way of negation. This idea seems to be more native to the oriental than to the occidental mind. The Chinese, for instance, excel in their recognition of the positive value of empty spaces in pictures,[4] and one of their poets pointed, long ago, to the essential significance, in the works of man, of such negative features as the hollow of a clay vessel, or the central vacancy of the hub of a wheel.[5]

When the biologist, having reached the preliminary formulation of his problem, turns to the second stage—the acquiring of

[1] Gibbon, E. (1896); see autobiographical and journal entries between 14 April 1761, and 15 October 1764, pp. 193 *et seq.* and 302.
[2] Masson, D. (1874), vol. I, p. 44.
[3] Cf. Poincaré, H. (1908), p. 307; Scott, G. (1914), p. 163.
[4] Binyon, L., in Fry, R., and others (1935), p. 12.
[5] Waley, A. D. (1934), p. 155.

data which may give a clue to the answer to his question—he embarks upon a troubled sea. To arrive at new facts, or even to direct some new light on to those already recorded, is, in itself, a task of extreme difficulty. Heraclitus, some five hundred years before the birth of Christ, said that "Nature loves to hide".[1] This is even more conspicuously true to-day than it was then, for, now that so many aspects of plant and animal life have already been induced to reveal themselves, the utmost refinements of methods and technique are often needed before anything fresh can be elicited. The work of the biologist does indeed require a many-sided equipment, since it demands both a talent for delicate manipulation, and for the technician's skills, in addition to the power of abstract thought—capacities which are rarely developed to the full in one personality. There is, however, so close an association between fingers and brain, that to circumvent this difficulty by handing over either the technical or the interpretative side of the research to another worker cannot but mean a serious loss of integration.

Apart from the internal hindrances, on a few of which we have just touched, there are often external obstacles, formidable enough to induce despair of ever arriving at the facts. When we think, for instance, of the imperfect state of the evidence for the past history of living creatures—all this evidence depending upon relatively rare and sporadic fossil remains, often consisting merely of detached and mutilated bits and scraps of debris—it seems a miracle that palaeontologists should have succeeded in acquiring all the detailed knowledge that is now available. Their success awakens one to the vast treasure-trove waiting to be uncovered by intensive study of even the most unpromising records. The difficulty of fragmentary evidence, and its triumphant conquest, are not peculiar to biology. A perfectly comparable situation arises in studying, for instance, the pre-Socratic philosophers. Such work has given results of the utmost significance, despite the fact that it is based, necessarily, upon evidence as fragmentary, and often as poorly preserved, as that

---

[1] Bywater, I. (1877), frag. x, p. 4, φύσις κρύπτεσθαι φιλεῖ; see also Patrick, G. T. W. (1888), p. 683, or (1889), p. 124, and Burnet, J. (1920), p. 133.

of the palaeontologist. Of the actual writings of Anaximander, for example, apparently only a single sentence has survived, transmitted at second-hand, but it has been described by a scholar in this field as "a jewel precious beyond estimate".[1] The biologist may perhaps be permitted to amuse himself with the fancy that the whole, to which such a sentence affords a clue, may be compared to an extinct plant or animal, the nature of which has to be deduced from a single fossilized scrap of the dismembered organism—a fragment teeming, however, with as much significance for him, as this sentence of Anaximander for a student of the pre-Socratics.

All direct evidence concerning biological questions depends upon the data the mind obtains from sense impressions. These data are greatly increased in scope and accuracy by technical advances, which not only have their obvious direct bearing, but also produce indirect results of the most essential kind. They often reorientate the mind towards natural things; this re-orientation in its turn educates the senses, which are far more dependent upon the mind than is sometimes assumed. In the seventeenth century, Henry Power,[2] in the first flush of enthusiasm for the microscope, wrote of the "minute Bodies and smallest sort of Creatures about us", as having been in the past "but slightly and perfunctorily described, as being the disregarded pieces and huslement of the Creation"; but when, with the aid of magnification, it was discovered, "how curiously the minutest things of the world are wrought", the mind became aware of unplumbed depths of complexity in organisms, and this awareness stimulated the eye to do its work more effectually than ever before.

The value of continually advancing technique is inestimable, so long as it is not allowed to become an end in itself, and thus to foster delusive industry of a pointless kind. The mechanical pleasure, for example, of cutting endless microtome sections, may lull the mind into serene inaction and comfortable passivity in regard to the problem to be solved. Equivalent side-tracking has to be guarded against in fields other than science. It was

[1] Jaeger, W. (1947), p. 34.   [2] Power, H. (1664), Preface.

painting that Sir Joshua Reynolds[1] had primarily in mind when he wrote of "A provision of endless apparatus, a bustle of infinite enquiry... employed to evade and shuffle off real labour, —the real labour of thinking".

Another difficulty of a slightly different kind arises out of the biologist's tendency to concentrate his efforts on a limited region of the 'not known', just outside the boundary of the 'known'. The result is that, when he succeeds in putting this unexplored region on to the map of the 'known', he is apt to take a disproportionate view of its import, in relation to that of the whole known field, of which it is, in fact, merely an insignificant marginal area. He might be compared to an antiquary, who, having in a remote corner of a vast cathedral unearthed a fragment of early masonry, invests his 'find' with an historic value out of all proportion to that of the corpus of previous knowledge about the building.

For convenience in handling, we have hitherto referred to the researcher's problem as if it could be considered in isolation; but, in so doing, we have oversimplified the issue. There is no such thing as *a* problem, just as there is no such thing as *a* fact or *a* thought. The questions which the biologist puts to Nature are integral parts of that living reticulum which is the intellectual life of the time; they cannot be seen truly except on the background of that reticulum, from which their vitality is drawn.

Furthermore, we have not only been thinking of problems in too isolated a fashion, but we have also given the word too narrow a connotation. It is true that the problems of a man's apprentice years will probably be straightforward questions, to which answers in familiar terms can be found; but the problems of his maturity should be something more than this. They may well involve the reconstitution of the frame of reference to which his specific questions are orientated, rather than any access of new knowledge; and the result may be a change of standpoint, from which fresh vistas, demanding exploration, are brought into view. Such work then falls into a category comparable, for instance, with that of Lavoisier—the great chemist, who also

[1] Reynolds, Sir J. (1797), Discourse XII, 10 December 1784, vol. I, p. 247.

advanced biology by throwing light upon the respiratory process; he is said to have instilled a new spirit into his science,
without ever having discovered a new substance or a new
phenomenon.[1] Such work in its method approaches the philosophical disciplines, in which the area studied remains broadly
the same from generation to generation, while all that changes
is the point of view. To the biologist, who is too often dis-
tracted from mental work by the insistent demands of manual
technique—from which the philosopher is free—such *general*
reorientation of thought is as difficult as it is rare. *Special* re-
orientation, related to one particular problem, is, however,
a common experience. Hitherto we have spoken as if a question,
once selected, remained constant to itself; but this is by no means
always the case. During the hunt for a solution, the problem
changes under the worker's hand, and may altogether outgrow
its first formulation. It is, indeed, one of the marks of a man born
for research, that his problems do not remain static, but come to
life, and quietly themselves assume the direction of his work.

Seen as a series, the problems, which any biologist sets before
himself, must be considered not only in connexion with the
thought of the time, but also in relation to his own individual
development. If the personality is fully integrated, these problems are not a sporadic group of discrete efforts, but form a
sequence showing a transition from juvenile to mature phases,
as continuous as the developmental history of a plant or animal.
Even when this analogy appears to break down, and the worker's
output, especially in his youth, seems to be an erratic and dis-
connected jumble, an onlooker of sufficiently penetrating insight
might detect, underlying these superficial vagaries, a movement, however circuitous, towards the eventual goal. This is
even more recognizable in the work of philosophers than of men
of science. For instance, a writer, who has made a special study
of Leibniz's intellectual life in his earlier years,[2] has come to the
conclusion that his whole system is traceable to five fundamental
concepts, all of which had already found distinct expression in
the work which he achieved as quite a young man. In the case

[1] Nordenskiöld, N. E. (1950), p. 265.　　[2] Kabitz, W. (1909), pp. 3, 127–34.

of Kant, again, prolonged and intensive study shows that, in spite of apparent inconsistencies and discrepancies in his successive writings, there is, in fact, an unbroken continuity in the movement of his thought as a whole.[1] The same continuous progression of the individual mind is seen still more clearly in the field of art. In poetry, for instance, though literary taste of the more primitive kind appraises a writer by those few works which are supposed to be his masterpieces, and have found their way into anthologies, serious critics treat a man's poetic output as an organic whole, and each individual work is assigned its place in a meticulously traced chronological sequence, which is often unexpectedly significant.

Few people are fully aware of the stages of their own mental life while they are passing through them; and it is only in the latter part of his career, if ever, that the biologist himself is likely to recognize the relation which the different phases of his work have borne to one another—a relation which depends upon his individual outlook, and is stamped with the impress of his whole life and thought. Indeed, in the choice of his successive problems— unless he has an unusual gift for introspection—he is wise to trust that his instinctive urge will point him to the road that is, at that moment, the right one for him, rather than to attempt a self-conscious analysis of his own past in relation to his possible future. Such analysis is apt to be too much like tearing up a seedling, and anatomizing its damaged roots, in order to encourage it to grow. As Meister Eckhart said, more than six hundred years ago: "A man ought not to work for any why...but only for that which is his being, his very life within him."[2] If the researcher relies upon his inner intelligence, it will—deploying itself beneath, or perhaps above, the level of his actual awareness—so direct him that an organic pattern, to which his production as a whole has conformed, will eventually come into view.

---

[1] Caird, E. (1889), vol. I, p. ix.

[2] Pfeiffer, F. (1924, 2nd imp. 1949), J. Eckhart's Serm. LXV, p. 163. It must be noted, however, that some of the writings that pass under Eckhart's name are now held to be of doubtful authenticity; see Clark, J. M. (1949).

CHAPTER II

THE MODE OF DISCOVERY
IN BIOLOGY

HITHERTO we have spoken about the solution of problems in an externalized way. It remains to see whether we can get any nearer the heart of the matter by attempting to examine what the process of discovery—the third stage in our sequence—actually means within the biologist's mind. It is true that, in additions to purely factual knowledge, the mental aspect is relatively unimportant; the limiting case would be, for instance, the finding of a plant or animal new to science by someone who was unprepared for anything of the kind and who brings home his *trouvaille* almost accidentally. At the other extreme, in interpretative research, mental activity obviously must predominate.

The whole question of the nature of biological discovery is merely one facet of a much more extensive subject, that of the provenance of 'creative' work in philosophy, literature, and the arts, as well as in science. Data concerning this question have been assembled on a considerable scale,[1] chiefly on the basis of a rather uncritical acceptance of anything which those who have achieved 'original' work of any kind have vouchsafed to disclose about how they did it. A study of such records leaves one feeling, however, that the people who are most vocally informative about their own procedure, are often just those for whose work the expression 'creative' is a mere courtesy title. Moreover, even where a stricture of this kind does not apply, such autobiographical statements can be accepted only with reservations. It is doubtful whether the most conscientious attempts of any man to analyse the thought-processes involved in the more original parts of his own work, are likely to be fully authoritative.[2]

[1] See for instance, Montmasson, J.-M. (1931), and Harding, R. E. M. (1948).
[2] Cf. Meyerson, É. (1930), p. 387.

It seems, as a rule, to be beyond the capacity of the man himself to compel into his consciousness the events that, in the normal course, pass unperceived in the hidden penetralia of his own mind; he can observe only their consequences, so far as they emerge into awareness, and at their grounds he merely guesses. The most we can hope is that his description may give us faint hints and glimpses of what has actually happened.

As regards scientific work, the classic statement of how discovery is conceived and born in the mind is still that of Henri Poincaré.[1] While recognizing the intrinsic suggestive power of Poincaré's lucid account, one may, however, regret that there has been a tendency to accord to it a more universal status than he himself claimed for it. His description was derived from what he observed retrospectively of the genesis of his own mathematical discoveries. When he looked back upon the paths by which his results had been reached, he was struck by a certain general parallelism in the course of events. He noticed, broadly, that in each example there was first a period (or periods) of intensive conscious work, which apparently failed to lead to any conclusion. Secondly, there was a *change* from conscious work. This change might take various forms: he instances either simple rest, or else a variation of place or pursuit, diverting the attention for a time to other matters. It was after such a period, in which the subject on which his mind had been bent was dismissed temporarily from the field of consciousness, that a third stage—in which the question he had been pondering passed from darkness into full light—suddenly supervened. This illumination put him into possession of a 'roughed out' solution of his problem which carried with it, at least for the time being, an irresistible conviction of truth—*une entière certitude*. This certainty did not, however, invariably survive critical examination, and, to put the whole thing on a sound basis, a fourth stage was required—another period of conscious work, in which the conclusions

---

[1] Poincaré, H. (1908). For an illuminating study of mathematical discovery, see also Hadamard, J. (1945). This book came into the writer's hands only after the present chapter had been completed; it should be consulted especially for a further account, with references, of the views of Poincaré and other more recent writers.

reached were studied, exactly formulated, and subjected to methodical scrutiny and testing.

When we review these stages, we see that there is nothing unexpected about them, except the fact that a phase characterized by the absence of conscious work precedes the final illumination, which gives birth to the hypothesis. Poincaré's own explanation of the whole process is that the essential factor is the combination—partly unconscious—of ideas whose capacity for forming a fertile union had not previously been recognized. Something of this rather obvious notion has been recurrent in the history of thought. Sir Joshua Reynolds, for instance, writing on art in the eighteenth century, declared that "Invention, strictly speaking, is little more than a new combination of those images which have been previously gathered and deposited in the memory".[1] Poincaré's special contribution lay in hinting at a type of mechanism which might implement this process. He supposed that the elements in the mind, through which discoveries may come, are like the hooked atoms of Epicurus; but he warns the reader that this comparison is *bien grossière*. These elements, when the mind is completely at rest, are, so to speak, attached to the wall. The first period of conscious work results in freeing and setting in motion a number of such atoms, selected as possible components of future unions. The conscious mind makes various tentative trials with these mobilized atoms, and induces them to form a variety of contacts and attachments, but these experiments are resultless. When the effort is given up, and the conscious mind is put to sleep, or else the attention is turned elsewhere, the atoms do not immediately revert to quiescence. They continue their unresting dance, in an automatic fashion, and they may—while consciousness is not, as it were, looking on—form innumerable further combinations at haphazard. The majority of these never gain entrance to the conscious mind at all; it becomes aware only of those which succeed in passing the scrutiny of what Poincaré calls the *sensibilité esthétique*—that sense of harmony and elegance, which fills so conspicuous a role in the work of the mathematician.

[1] Reynolds, Sir J. (1797), Discourse II, 11 December 1769, vol. I, p. 20.

According to Poincaré, it is only from amongst the combinations which appeal to this special sensibility, that the solution of the problem can arise.

It is dangerous to accept unreservedly this picture of the actual 'how' of discovery, dependent as it is upon the postulated performances of a metaphorical apparatus of 'hooked atoms'—an invention which Sir Isaac Newton stigmatized in another connexion as "begging the Question".[1] Poincaré's attempt to provide visual images of mental processes, which are, in fact, non-picturable, cannot have more than a suggestive value, but—apart from this—we have to consider how far the external descriptive features of his account are applicable to discovery in fields outside mathematics. Biologists, as well as mathematicians, would no doubt agree that successful hypotheses can only come into being after a preparatory period of conscious intellectual work. A great teacher of ancient China was voicing something cognate when he spoke of "the long expenditure of strength, and then one day, in a flash, everything becoming linked up together".[2] In those branches of biology, however, which deal primarily with form and structure, it is doubtful whether the succeeding phase, which Poincaré describes, is at all characteristic—the phase, that is, in which withdrawal of attention from the subject precedes the advent of the hypothesis. This difference is probably due to the fact that, in morphological work, visual impressions, immediate or remembered, play a part which cannot be paralleled in the more abstract quantitative disciplines.

When a biologist looks to his own experience, instead of trying to apply, in his very different field, the interpretation which the mathematician is justified in using in his own subject, he may find that he can either confirm or refute a suggestion as to the origin of biological discovery, which seems to the present writer to meet the case. It is that new hypotheses come into the mind most freely when discursive reasoning (including its visual component) has been raised by intense effort to a level at which

---

[1] Newton, Sir I. (1931; reprint of 4th ed. 1730), bk. III, quest. 31, p. 388.
[2] Chu Hsi of the Sung Era, quoted in Fung Yu-lan (1947), p. 216.

it finds itself united indissolubly with feeling and emotion. When reason and intuition attain to this collaboration, the unity into which they merge appears to possess a creative power which was denied to either singly. It is not possible to offer strictly scientific evidence for the idea that not only reason but emotion has a function in biological discovery, as it admittedly has in creative work in the arts; we can only point to slight indications which are at least compatible with its truth. It is recognized, for instance, that the moment at which a fruitful combination of ideas enters the awareness, is often charged with a peculiar feeling of joy,[1] which precedes and seems independent of, the rational satisfaction of goal-attainment. Such instantaneous delight is fully congruent with Spinoza's definition of pleasure as "the passion. by which the mind passes to a higher state of perfection".[2] This emotional element in discovery is perhaps the factor which makes it so elusive, and so refractory to organized control.

[1] Cf., for instance, Varendonck, J. (1921), p. 282; Wallas, G. (1926), p. 97.
[2] [Spinoza, B. de] 'B.D.S.' (1677), *Ethices*, pars III, prop. XI, Scholium, p. 105: "Per *Laetitiam*...intelligam *passionem, quâ Mens ad majorem perfectionem transit.*"

CHAPTER III

THE LOGICAL BACKGROUND OF THE
BIOLOGIST'S PROBLEM

IN the preceding chapter we touched upon the very core of problem-solution—the actual process of discovery, or, in other words, the origin of hypotheses. We have now to turn to the fourth stage of the biologist's work—the testing of his interpretation. This means that we must look at his activities afresh, on the background of logical thought, and inquire in what way, and to what extent, they stand related to it.

The first question, to which the biologist needs an answer, is: What, exactly, is meant by *logic*? The broader definitions—culminating in such a view as Joachim's, that it is "the science of knowledge-or-truth"[1]—cover such an extensive field that logic ceases to be distinguishable from certain other philo-sophical disciplines.[2] On the other hand, Kant's conception[3]— that logic is limited precisely to the formal rules of all thought, and has no concern with the *content* of knowledge—is somewhat unsatisfying. For the scientific worker, definitions that may make a stronger appeal are that logic is "the reflection of know-ledge upon itself",[4] or that it is "the science of implications", and that its "one aim...is to make our meaning clear, both with regard to what is meant and to what that meaning implies".[5] There has been, indeed, a tendency to revolt against the chain-like character of the reasoning sanctioned by rigid logic, as

[1] Joachim, H. H. (1948), p. 20; for another, less exaggeratedly broad definition, see Johnson, W. E. (1921–4), pt. I, 1921, p. xiii.
[2] Walsh, W. H. (1947), p. 5, footnote.
[3] Kant, I. (1902–38). Bd. III, 1904, *Kritik der reinen Vernunft*, 2nd ed., 1787, *Vorrede*, p. 8, "die Grenze der Logik aber ist dadurch genau bestimmt, dass sie eine Wissenschaft ist, welche nichts als die formalen Regeln alles Denkens... ausführlich darlegt und strenge beweiset." Translation, Smith, N. Kemp (1933), p. 18; see also *Einleitung*, p. 80: "Also ist das bloss logische Kriterium der Wahrheit...zwar die *conditio sine qua non*, mithin die negative Bedingung aller Wahrheit: weiter aber kann die Logik nicht gehen." Translation, Smith, N. Kemp (1933), p. 98.
[4] Bosanquet, B. (1911), vol. I, p. 3.   [5] Mackenzie, J. S. (1917), pp. 102–3.

22

confining the mind too narrowly to the conventional linear concept of thought.[1] In the view of certain idealist philosophers, the logician's concern is not with isolated chains of reasoning, but with the detection of the relations of any special problem to the surrounding nexus, and thence, ultimately, to the whole of reality.[2]

With so much difference of opinion among experts, the student of biology can hardly be expected to adopt, blindly, the rules of text-book logic in his own procedure; but it may help him, in testing his conclusions, to analyse the methods, which he himself has used, in their relation to those general logical principles on which there is fair agreement. It is true that the biologist's results have, in fact, seldom been *attained* by the use of those rules with which he afterwards finds that his reasoning harmonizes; he may, indeed, have been quite unaware that such rules had been formulated, and yet his argument, when examined, may fall readily into place in the framework which they provide. Though it is true that the biologist can no more penetrate the mystery of discovery by the help of logical rules, than the painter can learn how to create a picture from the art critic's painstaking analyses, yet, in research, the logical framework is not without its underlying significance. It is fully accepted that manipulative techniques must be *learned*, but the idea that we also have to *learn* how to think effectively is somewhat alien to the scientific temperament. Though the biologist, in his undergraduate days, has, to some extent, to submit to this educational process, much of its value is obscured for him by the fact that he passes through it unwittingly. The upshot of such unconscious study is liable to be patchy and inadequate; its result sometimes suggests that the conscious discipline of a training in logic might have had the effect of clarifying the biologist's thought, which is liable to be clogged by those elementary confusions which the logical scalpel could have removed with ease. One such confusion,[3] which at times vitiates biological argument, is that of asserting the

---

[1] See Chapter IV, pp. 45, 46, of the present book.
[2] Bosanquet, B. (1920), and Metz, R. (1938; reimpression, 1950), pp. 349–50.
[3] Johnson, W. E. (1921–4), pt. I, 1921, ch. V, pp. 54 *et seq.*, and Hobhouse, L. T. (1896), p. 409. Cf. also Woodger, J. H. (1929), p. 322.

antecedent on the ground of the consequent. This comes about because it is not always realized that it is far more difficult to deduce a cause from an effect, than to see what effect a particular cause will produce; for the cause to which we ascribe an effect may, indeed, be one of a number of *possible* causes, but not necessarily the one that actually operated. For instance, we might claim with justice that *if* Natural Selection has governed the past development of the organic world, animals and plants to-day will show adaptations to their environments. Further we may observe, rightly, that animals and plants *are* adapted to their environments. It is not justifiable, however, to conclude from these two statements that Natural Selection has, as a matter of history, controlled the development of the organic realm. All we can say is that Natural Selection is one of antecedent factors which *might* have produced this effect; but our reasoning gives no guarantee that this effect has not been, in fact, brought about by other causes, perhaps of a totally different kind. St Thomas Aquinas, in the thirteenth century, when discussing astronomy, recognized a corresponding logical crux. He pointed out that the notion of eccentrics and epicycles was accepted because it accounted for the observed movements of the heavenly bodies, but he considered that this did not amount to proof, since it was possible that some different hypothesis might serve equally to save the phenomena.[1] He is referring here to the familiar Greek phrase, σῴζειν τὰ φαινόμενα,[2] which Milton translated, "To save appearances", when he put it into the mouth of the Archangel Raphael in his discourse to Adam of astronomical theories.[3] *Saving the phenomena* may be regarded as a key expression, indicating a certain difference, rather in relative stress than in method, between classical and modern science.[4] The scientist to-day starts with the phenomena, and, step by step, arrives at his theoretical interpretation. The Greeks of Plato's period, on the

---

[1] Aquinas, St Thomas, in Migne, J. P. (1845, etc.), vol. 1, p. 741, *Summa Theol.* 1, quest. 32, art. 1; "Sicut in astrologiâ ponitur ratio excentricorum et epicyclorum, ex hoc quòd hâc positione factâ possunt salvari apparentia sensibilia circa motus coelestes." Translation in Aquinas, St Thomas (1911, etc.), 1, no. 2, pp. 60–1. For a further account of Aquinas's views on hypothesis, see Brodrick, J. (1928), vol. II, pp. 330–1.   [2] Burnet, J. (1914), p. 11.
[3] *Paradise Lost*, VIII, 81.   [4] Wightman, W. P. D. (1950), p. 35.

other hand, paid attention primarily to theory, and only secondarily to facts. They tended to begin with a generalized idea; if they found that the particulars were not incompatible with this idea, they considered the phenomena to be 'saved'. At the present day we realize, in a way that the Greeks could not have done, how indescribably complex, and difficult to interpret, the appearances are; and we are able to see that an individual hypothesis can hope only to 'save' a certain group, in the sense of demonstrating its coherence; from this group, by the process of *induction*, some further generalization may then be evolved. This method is Aristotle's ἐπαγωγή, "the approach from particulars to the universal".[1] It may be regarded as, in the strict sense, scientific, since it is a common, shareable process,[2] while 'discovery', which is typically the intuitive work of a single mind, is akin to art rather than to science. Induction takes its start from an individual fact, observed in some particular case. For example, it might be noticed that, in the petiole of the peltate (shield-shaped) leaf-blade of a garden nasturtium (*Tropaeolum*),[3] the vascular skeleton is approximately radial (stem-like). Further observations would show that this is true, not only of other leaves of the same plant, but also of any leaf belonging to the same species, of which one chose to cut sections. Furthermore, petiole structure on an identical plan is also to be found in peltate leaves belonging to other families. The scrutiny might, moreover, be extended to the peltate leaves which sometimes occur as an abnormality in families in which the leaves are ordinarily non-peltate; these exceptional peltations will be found to show a similar construction. Observations of this kind, if they could be distributed over a wide enough field, lead to the inductive conclusion that, in any example, as yet unexamined, a peltate leaf-blade would be found to be borne on a petiole with radial anatomy. Such inductive reasoning from observations, as Sir Isaac Newton long ago pointed out, is "no Demonstration of general Conclusions; yet it is the best way of arguing which the

---

[1] Cornford, F. M. (1935), p. 185.
[2] This view of the nature of science is expounded, for instance, in Dingle, H. (1931).
[3] Cf., for instance, *T. majus* L., Troll, W. (1935, etc.), vol. I, p. 1205.

Nature of Things admits of".[1] Inductive reasoning, at the best, can never, indeed, lead to certainty; a high probability is the utmost it can offer.[2] It can, however, be tested to a limited extent by application of the other primary procedure of logical reasoning—deduction. This, in contrast to induction, is essentially a movement from the general to the particular. In the classification of the living world, for instance, the process of arranging organisms into successively more inclusive groups (e.g. individuals into species, species into genera, and so on) may be called inductive; it works from below upwards. On the other hand, if we begin with a large synthetic group, and analyse it downwards, into smaller groups, we are proceeding deductively.[3]

In its logical form, deduction is typified by the syllogism. If, in our botanical example concerning peltate leaves, we use the conclusion, which we have reached inductively, as the major premiss of a syllogism, we can deduce results from it which are capable of being tested by observation. Our inductive conclusion was that *the petioles of peltate leaves have a radial anatomy.* Treating this as the major premiss, we may then choose a number of species, *A, B, C,* etc., with peltate leaves (not included among those from which the evidence for the original induction was drawn), and use the statement, *the leaves of A, B, C, etc., are peltate,* as the minor premiss. From these two statements we deduce the conclusion that *the petioles of A, B, C, etc., will show radial anatomy.* If we afterwards find by actual observation that this is true, it affords, as far as it goes, evidence for the probability of the statement to which induction led us. All this argument does not, however, give the biologist much help, for the procedure we have outlined is precisely that which common sense dictates to the investigator, and which he uses as a matter of course,

---

[1] Newton, Sir I. (1931; reprint of 4th ed., 1730), bk. III, quest. 31, p. 404.
[2] On induction and probability, see Keynes, J. M. (Lord) (1921), Williams, D. (1947), and, for detailed and highly technical treatments, Wright, G. H. von (1951) and Wisdom, J. O. (1952). From the standpoint of the biological sciences, special interest attaches to Kneale, W. (1949), from which (cf. pp. 22 and 214) it becomes clear that the type of probability which concerns the biologist must be distinguished from the measurable probability which the mathematician studies in connexion with chance.
[3] On synthesis and analysis in classification, cf. Zimmermann, W. (1930), p. 5.

without ever thinking of it under the grandiloquent terminology of logic. For the same reason, the various more elaborate techniques, which logicians recommend, are apt to offer little to the biologist beyond a nomenclature and a detailed discrimination of the tools with which his hand is already familiar. We might, for instance, apply Mill's *Method of Agreement* and *Method of Difference*[1] in order to arrive at a general statement about dorsiventrality in a leafy shoot. We might either compare among themselves those outgrowths from shoots, which are essentially dorsiventral (i.e. leaves)—*Method of Agreement*; or we might compare these dorsiventral outgrowths with others that are essentially radial (i.e. lateral shoots)—*Method of Difference*. This is undoubtedly a sound procedure, but it is just what any observer, who was seeking for light on dorsiventrality in shoot outgrowths, would adopt, even if it had never occurred to him to direct his thinking according to the rules of logic. The biologist, indeed, cannot but feel that a self-consciously logical approach would often hinder rather than promote his work. It is true that inductive reasoning, tested deductively if possible, is the means in science whereby the passage from observation to a generalized statement, of some degree of probability, is achieved; but further than this induction cannot go.[2] Even Bacon, notwithstanding his fervent advocacy of the inductive method, says that, after the "*Presentation* to the *Intellect* of all known *Instances*...we think it useful to grant permission to the Intellect...to gird itself up and attempt the *Interpretation of Nature* in the Affirmative".[3] This last phrase may well point, as Kneale[4] has suggested, to the method of discovery in natural science which involves hypotheses. Induction by itself can afford no clue to the *significance* of the generalizations which it achieves. Taking the example, which we have already been tracing, concerning the association of a peltate leaf-blade with a radial petiole, we find that we cannot

---

[1] Mill, J. S. (1843), vol. I, p. 450. Duns Scotus distinguished these two methods of induction 650 years before Mill; see Carré, M. H. (1949), p. 147.

[2] Wisdom, J. O. (1952) came into the writer's hands after the present chapter was completed; in this book Wisdom states his conviction that "induction plays no part whatever in science" (Preface, p. vii).

[3] Bacon, F. (1620), *Novum Organum*, bk. II, aphor. 11 and 20; for translation, Kitchin, G. W. (1855), pp. 125, 155.   [4] Kneale, W. (1949), p. 53.

understand its inwardness except through the process of intuitive discovery. It is this process which resulted in the hypothesis that leaves are 'partial-shoots', with a tendency to develop 'whole-shoot' characters.[1] An example of this urge is seen in the peltate leaf, for a leaf of this type shows a radiality of form which recalls the 'all round' symmetry of the shoot, rather than the dorsiventrality of an ordinary leaf. This trend is not only revealed externally in the shape, but also internally in the radial anatomy of the petiole. The hypothesis in question thus links together facts, which, if viewed on inductive lines alone, seem unrelated. We may say, in general, that induction establishes the probability that certain features will be found associated, but it is not concerned with the *raison d'être* of this association. An hypothesis[2] has a contrasting function, since it aims at disclosing the underlying significance of the factual associations certified by induction. In other words, the hypothesis resolves a statement, which induction has left as a mere factual juxtaposition—inorganic, and thus incapable of further growth—into an organized living whole. It is clear that such a claim as this can be maintained only for those hypotheses which will endure critical scrutiny—and such hypotheses are relatively rare. Though the modern investigator uses less picturesque forms of expression, he may sympathize with Robert Boyle's comparison of the "noble Experiments" of scientists, to the wealth of "Gold and Silver and Ivory", which King Solomon's navigators carried from Tarshish; and he may also share Boyle's regret that the explanations offered, often show, on the other hand, a triviality analogous to that of the rest of the cargo, since they are "Theories which either like Peacocks feathers make a great shew, but are neither solid nor useful; or else like Apes, if they have some appearance of being rational, are blemish'd with some absurdity or other".[3] To-day, as much as in the seventeenth century, a method is needed for distinguishing between hypotheses of lasting value, and Solomon's apes and peacocks, with their undoubted but delusive charm. In the physico-chemical

[1] Cf. Arber, A. (1950), chs. VI–VIII.
[2] On the historic origin of the hypothesis concept, see Robinson, R. (1941).
[3] Boyle, The Hon. R. (1680), pp. 433–4.

sciences, and in the type of biology which shares the technique of these disciplines, valid hypotheses can often be distinguished from those that are inadequate by the test of quantitative confirmation. Newton's dictum that the "main Business of natural Philosophy is to argue from Phaenomena without feigning Hypotheses",[1] does not mean that he rejected hypotheses in general, but that he was unwilling to admit any which were not susceptible of experimental proof.[2] In the region of biology most remote from physics and chemistry, however, the very nature of the hypotheses employed prevents their being tested on any mathematical-experimental system, such as Newton had in mind. Such biological hypotheses represent *ways of looking at Nature*, so that, like artistic creations, they cannot be assessed by any simple and easily defined procedure. They resemble the interpretations which an artist sets down, which are disengaged from his visual impressions without any *conscious* 'discursive' thought.[3] As Plotinus said long ago, "the artist reasons only when he is at a loss; when all goes smoothly, his art is controlling him and doing his work".[4] Biological hypotheses of the more creative kind have, like the painter's interpretations, to be estimated by indirect methods. The key to such testing is the fact that an inadequate or erroneous hypothesis leads, as it were, an isolated existence, and is united imperfectly, if at all, to a surrounding nexus of thought. In terms of the 'coherence' theory of truth,[5] we may say that, when an hypothesis, which offers a solution of a particular problem, has been reached, its status must remain uncertain until it has revealed a capacity for extending in all directions, everywhere showing natural and inevitable connexions with valid interpretations of other parts of reality. It can then be visualized as "conformable to the nature of things" (κατὰ φύσιν).[6] When any hypothesis has thus

[1] Newton, Sir I. (1931; reprint of 4th ed., 1730), bk. III, quest. 28, p. 369.
[2] Cf. Andrade, E. N. da C. (1947), p. 13; see also Wright, G. H. von (1951), pp. 22–3.
[3] 'Discursive thought' passes, by a rigidly logical process, from premiss to conclusion. It is Hamlet's "discourse of reason" (I. ii) which Milton's Raphael contrasts with intuition (*Paradise Lost*, v. 488).
[4] Dodds, E. R. (1923), Plotinus, *Enn.* IV. iii, 8, p. 90.
[5] See Chapter VI, pp. 70 *et seq.*, of the present book.
[6] Whittaker, Sir E. T. (1946), p. 57.

justified its place, it has achieved graduation into a *theory*, which may be defined as a broad view or insight, relating phenomena comprehensively. When a theory is well established, and can claim a high degree of generality, it is often called a 'law'; but, at least in the biological field, it is doubtful whether the bestowal of this title can be justified. Originally the expression, 'laws of nature', when used in science, referred to those direct edicts of the Almighty, which were held to control material things; in this sense there was a close analogy with human law, so that the term was fully applicable. In modern writing, however, in which a 'law' of nature stands for a "theoretical principle deduced from particular facts",[1] the word 'law', which suggests compulsion, is obviously out of place.

If we turn from the problem-solutions that succeed to those that fail, we find that the path of thought is littered with hypotheses that have either been functionless from the first, or have served for a time, and then have yielded place to some new generalization. It is often a misfortune that such hypotheses are liable to be swept out of sight by the instinctive tidiness of systematic thinking, on the assumption that to pay attention to them would be a waste of time and energy. This assumption is a mistake, for the truth or falsity of hypotheses is not absolute, but relative, and even those that fail to find acceptance may yet have their own value. Orientation to a question may be markedly changed by the review of a number of suggested solutions; even if they all have to be rejected as inadequate, they may be able to contribute some elements of truth to a later hypothesis. Much is lost if we attend only to the currents which have led straight on to what we are pleased to regard as the enlightened conceptions of our own period. If we look carefully at hypotheses discarded in the past, we find that, like dormant buds, they are sometimes capable of initiating new branches of thought, after the vitality of the main trunk, which once overshadowed them, has become exhausted.

Even hypotheses which in the long run deservedly lose their status, may for a time have had their function in co-ordinating

[1] Cf. definition in *Oxford English Dictionary*.

facts. For instance, the hypothesis used by Claudius Ptolemy (second century A.D.), that the earth is the fixed centre of the universe, to us to-day is a misapprehension, but in the past it was found to 'work' with startling effectiveness; on the Ptolemaic system,[1] astronomers, until the sixteenth century, accounted with considerable success for the apparent motion of the heavenly bodies.

The acceptance of an hypothesis, and its resulting adoption as a theory, do not mean that it is 'proved'; and it is at least a possible view that it is to 'authenticate', or to 'justify',[2] rather than to 'prove', which should be the biologist's aim. These terms have their use in suggesting an approach towards verification, which is not necessarily empirical,[3] but which reveals the full implications of hypotheses, and exposes any inconsistency between themselves, or with thought in general.[4] In the purely biological, and thus autonomous, aspects of science, we are, strictly speaking, outside the pale of logic, and there can be no question of certainty or proof. We cannot even reach probabilities expressible in mathematical terms; we can look only for psychological probability.[5]

When we review, broadly, the relation of logic to the biologist's thought, we are left feeling that it can never be more than ancillary. As John of Salisbury recognized in the twelfth century, though logic can further other studies, it cannot by itself quicken the soul to yield fruit of philosophy.[6]

---

[1] For a brief summary of the Ptolemaic system, see Wightman, W. P. D. (1950), pp. 40–1; and Singer, C. (1941), p. 84, and Fig. 40, p. 87.
[2] Cf. Wright, G. H. von (1951), pp. 20 et seq.
[3] Cf. Urban, W. M. (1939), pp. 213 et seq.
[4] Cf. Stace, W. T. (1932), chap. XIII.
[5] Johnson, W. E. (1921–4), pt. I, 1921, p. xxxix.
[6] Poole, R. L. (1920), pp. 185–6.

CHAPTER IV

THE BIOLOGIST'S USE OF ANALOGY[1]

I N considering the logic of problem solving, we have thought about the use and limitations of induction as a scientific process, but we have still to consider the rights and wrongs of a cognate method of using inference from sampling[2]—the argument from analogy. This is a topic which should not be approached on narrowly specialist lines; it cannot be discussed effectively except in relation to thinking in general. Everyone agrees that, in our mental workshop, one of the tools, which is most constantly employed, is the perception of likeness. This tool, indeed, is one for which we often have an undue craving, because of the degree of emotional satisfaction which its use affords. The "pleasure which the mind derives from the perception of similitude in dissimilitude", was stressed by Wordsworth[3] in the Preface to the *Lyrical Ballads*, while, long before, Francis Bacon had warned the scientist that "The Human Intellect, from its peculiar nature, easily supposes a greater uniformity and equality in things than it really finds. . .it feigns parallels, correspondents, and relations which do not exist".[4] Nevertheless, despite these dangers of wishful thinking, we cannot reject the evidence of similitudes, for the claim that analogy is a staff of the mind, and that even mistaken analogies may be its stepping-stones,[5] is well founded. Bacon himself, though he was so convinced an exponent of strictly inductive methods, supplemented them by many suggestions derived from analogy.[6] He

---

[1] The substance of Arber, A. (1946b), is, by permission of the editor and publisher, incorporated in this chapter; see Acknowledgements, p. ix.

[2] See Whittaker, Sir E. T. (1946), p. 37.

[3] Wordsworth, W. (1800), p. xxxii; Wordsworth may have had in mind the "intuitive perception of the similarity in dissimilars", to which Aristotle refers; see Bywater, I. (1909), p. 71.

[4] Bacon, F. (1620), *Novum Organum*, bk. I, aphor. 45; for translation, see Kitchin, G. W. (1855), p. 21.

[5] Gregory, J. C. (1945), pp. 247–8; as an example of the value of analogies in scientific study, see Farber, E. (1950), which deals with their use in chemistry.

[6] For a study of Bacon's use of analogy, see Fischer, K. (1857), pp. 125–39.

urged inquiry into the subject, since he held that analogies unite nature, and are fundamental for the sciences.[1] In the eighteenth century, Berkeley stressed a corresponding view, holding that natural philosophers were distinguished from other men by "a greater largeness of comprehension, whereby analogies, harmonies, and agreements are discovered in the work of Nature".[2] Jevons, again, who was born 150 years later than Berkeley, was even more emphatic about the outstanding importance of similitudes. He wrote that "In every act of inference or scientific method we are engaged about a certain identity, sameness, similarity, likeness, resemblance, analogy, equivalence or equality apparent between two objects".[3] It is indeed true that thought cannot proceed without laying hold on similitudes. We arrive at general ideas by assembling similar phenomena, and reaching thence to concepts or 'universals', of which the phenomena are individual expressions. In scientific thought, especially, the perception of likeness and unlikeness is perhaps the most indispensable of all clues for unravelling problems of widely different types. No doctrine of probability, for example, can be established without its help; as Butler wrote, "That which chiefly constitutes Probability is expressed in the Word Likely, i.e. like some Truth, or true Event; like it, in itself, in its Evidence, in some more or fewer of its Circumstances. For when we determine a thing to be probably true...'tis from the Mind's remarking in it a Likeness to some other Event, which we have observed has come to pass."[4] Though Butler's special emphasis on the relation of similitude and probability was all his own, cognate ideas had been expressed long before. Plato set peculiar store by resemblance and the 'likely', holding, as he did, that the visible world is "a changing image or likeness (*eikon*) of an eternal model.... The inference is that no account that we or anyone else can give of it will ever be more than 'likely'."[5] This

[1] Bacon, F. (1620), *Novum Organum*, bk. II, aphor. 27; "Itaque convertenda planè est opera, ad inquirendas et notandas rerum Similitudines et Analoga... Illae enim sunt, quae Naturam uniunt, et constituere Scientias incipiunt."

[2] Berkeley, G. in Jessop, T. E. (1949), *Principles of Human Knowledge* (1st ed., 1710), pt. I, 105, p. 87.

[3] Jevons, W. S. (1877), 2nd ed., p. 1.

[4] Butler, J. (1736), Introduction, p. ii.   [5] Cornford, F. M. (1937), pp. 23–4.

implies an inevitable relativity in all accounts of the universe, and any man of science, whose sympathies are with Plato, will feel that he can be on sure ground only when his work takes the form of *comparative* thought within the framework of some 'likely' account, for which no absolute validity is claimed. In physics, a typical instance is summarized in D'Arcy Thompson's aphorism: "Newton did not shew the cause of the apple falling, but he shewed a similitude between the apple and the stars."[1] In biology, it is by consciously or unconsciously following Plato's suggestion that research workers have established those branches of the study which are in essence comparative. It has been claimed[2] that Vicq d'Azyr first employed the term 'comparative anatomy' in the eighteenth century, but the expression can in reality be traced much further back. Thomas Willis[3] used it in 1664 in his *Cerebri anatome*, while Nehemiah Grew employed it in relation to roots in his book of 1673,[4] and in 1681,[5] he applied it also to animal organs. As Charles Singer has pointed out, a serious disservice was done to science when—after Vesalius and other early workers had set human anatomy on a broad comparative basis—Spigelius made a reactionary separation between this study and that concerned with animal bodies, thus bringing down human anatomy to the level of a technical discipline.[6] The purely comparative aspect of the sciences of plant and animal form was emphasized in the days of Goethe and A. P. de Candolle, but fell somewhat into abeyance in the post-Darwinian period, when the passion for tracing phylogenies was at its height. In botany the comparative method has been rehabilitated in the twentieth century by the school of morphologists headed by Wilhelm Troll.[7]

Even the simplest pure description is basically comparative. This relativity may be concealed by the introduction of units of measurement, but, when traced to their origins, these prove to

[1] Thompson, Sir D'Arcy W. (1917), p. 6.
[2] Thompson, Sir D'Arcy W. (1942), p. 9, footnote.
[3] Willis, T. (1664), p. 66; cf. Schmid, G. (1935), p. 597.
[4] Grew, N. (1673), pt. II, p. 54, *The Comparative Anatomy of Roots*.
[5] Grew, N. (1681).
[6] Singer, C. (1931; 2nd ed. 1950); cf. chap. VI, pp. 202 *et seq.*, *Rise of Comparative Method*. [7] Cf. Troll, W. (1935, etc.)

be themselves comparative. The comparative method is indeed inescapable, but its value is apt to be underestimated, not because of any flaw inherent in its own structure, but because of the way in which it is liable to be abused. As the stranger from Elea says in Plato's *Sophist*, "a cautious man should above all be on his guard against resemblances; they are a very slippery sort of thing".[1] A frequent danger is the tendency for comparisons—especially those which ring true only under some special thought-aspect—to become mechanical clichés. The warning note struck in Shakespeare's sonnet, "My mistress' eyes are nothing like the sun", is still as necessary as it was when it was uttered, for conventional language continues to belie everything which it touches "with false compare". In science, similes which—so long as they were firmly welded with the thought that prompted them, had a certain merit, are apt to outlive that thought and sink into mere counters. Botanical text-books, for instant, are given to noting a routine parallel between a carpel and a foliage leaf, without any rigorous analysis of the content of this similitude, thus destroying such value as the comparison originally possessed.

A subtler and more insidious defect in the use of comparison arises out of the very structure of language. Running through both spoken and written thought, there is a linear time sequence, which gives a certain significance to the order of words. When we say, "The portrait is like the sitter", there is no doubt as to which term is the standard of comparison, but even if we merely say, "*A* is like *B*", the order of the words carries the implication that *B* is the standard to which the comparison is referred. This difficulty may be avoided to some extent by saying, "*A* and *B* are alike", instead of "*A* is like *B*". We need, indeed, to bear constantly in mind the distinction between irreversible comparisons, in which one term is definitely the standard, and reversible comparisons, which are applicable equally in either direction. This necessity is illustrated in the botanical simile already cited. If we say, "The carpel is like the foliage-leaf", the sequence of the sentence hints at a prejudgement in favour of

---

[1] Cornford, F. M. (1935), p. 180.

the hypothesis that the foliage-leaf is the original model to which the carpel conforms.

Modern scientists tend to shun the word 'explanation',[1] and to speak rather of "tracing relations between phenomena";[2] such tracing of relations carries us beyond mere similitude and resemblance to the further concept of analogy. The term analogy (ἀναλογία) has been used in every variety of signification, down to simple similitude, and it has hence become a tempting word for each man to define in his own way; but it seems best always to bear in mind its original meaning of *proportion*, which is also one of the meanings of λόγος. It is thus limited to cases in which resemblance of *relations*, rather than similarity, pure and simple, is the point to be emphasized. It is upon analogy, in this relational sense, that the whole of science has been built up. Since only a few of the countless phenomena in the universe can actually be observed, reliance must, in the last resort, be set upon the belief that the relations, which we are debarred from observing, are analogous to those in the field which is open to our perception. Without analogy, no wide generalizations could ever be formulated. Butler recognized this when he wrote: "It is then, but an exceeding little Way, and in but a very few Respects, that we can trace up the natural Course of things before us, to general Laws. And it is only from Analogy, that we conclude, the Whole of it to be capable of being reduced into them; only from our seeing, that Part is so. It is from our finding, that the Course of Nature, in some Respects and so far, goes on by general Laws, that we conclude this of the Whole."[3]

In the Middle Ages, the argument from analogy was not used merely in Butler's restrained sense, but it permeated the general thought of the period to an almost startling degree. It found its focal point in the theory that man himself could be understood by analogy with the scheme of things as a whole, and vice versa. Basing our statement on Meyer's definition,[4] we may say that this theory is the doctrine that man, in state, characteristics, and

[1] On explanation, see also Chapter v, pp. 58–9, of the present book.
[2] Dingle, H. (1937), p. 26.
[3] Butler, J. (1736), pt. II, chap. IV, sect. III, p. 189.
[4] Meyer, A. (1900), pp. 1, 2, etc.

conditions of existence, represents the state, characteristics, and conditions of the whole world; that he is a picture of the world; an epitome of the all; the world in miniature; the microcosm. On the other hand, the world-whole—the macrocosm—is to be conceived as a picture of man; an animated being equipped with a soul; a human organism writ large; a magnified man. Meyer contends that the expression 'microcosm and macrocosm' is inexact, and owes its currency to its alliteration; he maintains that, strictly speaking, the contrasting term to microcosm (small world) should be makranthropos (large man). It is doubtful, however, whether such a change in terminology would represent the medieval view accurately. It could be justified only if the comparison were held to be completely reversible, but in early thought it seems rather to be the macrocosm which was the primary idea—an idea which is imitated and illustrated in man.

When man is regarded as the microcosm, it is his sharply delimited finiteness which is stressed, not the innumerable connexions which make him part of the universe as a whole. A thinker who visualizes the universe as framed microscopically in an individual man, resembles an artist who translates his vision of a whole landscape in terms of a small rectangle of canvas. By thus isolating, artificially, a fragment of reality, the painter achieves a special microcosmic quality which would be lacking in a comprehensive panorama without defined boundaries.

The doctrine of the relation of man to the universe being that of microcosm to macrocosm, was far from new in the Middle Ages; it was foreshadowed in ancient China,[1] and among the Greeks[2] by the Pythagorean school, from which Plato adopted the idea of this "mimetic relationship between the individual and the universe of which he is a part".[3] In his myth of creation in the *Timaeus*, Plato expresses this relationship pictorially. He tells us that the Demiurge, in order to compound the immortal part of the souls of men, "turned once more to the same mixing

---

[1] Waley, A. D. (1939), p. 253.
[2] The history of the theory among the Greek, Arabic, and Jewish philosophers, and its relation to the thought of Spinoza, is traced in Wolfson, H. A. (1934), vol. II, pp. 7, 8, etc.
[3] Tredennick, H. (1933), Introduction, p. xxi.

bowl wherein he had mixed and blended the soul of the universe, and poured into it what was left of the former ingredients, blending them this time in somewhat the same way, only no longer so pure as before".[1] Thus, according to Plato, the relation of macrocosm to microcosm involves marked contrast as well as likeness; the macrocosm is everlasting, but in the microcosm it is the divine reason alone which is the imperishable element.[2]

Seneca (d. A.D. 65) may be cited as a later classical writer, whose view of the world was strongly influenced by the doctrine of the microcosm and macrocosm. In his *Quaestiones Naturales* he wrote: "My firm conviction is that the earth is organized by nature much after the plan of our bodies.... So exactly alike is the resemblance to our bodies in nature's formation of the earth, that our ancestors have spoken of veins (= springs) of water."[3]

Among those who, in the medieval period, made the doctrine of the macrocosm and microcosm their own, was the scholar-mystic, Hildegard of Bingen (1098–1180);[4] she developed and elaborated it, stamping it with the impress of her powerful personality. She saw permeating the universe—as the soul permeates the individual man—a single vital spirit, which she unveils in terms instinct with poetry: "I am that living and fiery essence of the divine substance that glows in the beauty of the fields. I shine in the water, I burn in the sun and the moon and the stars....I breathe in the verdure and in the flowers, and when the waters flow like living things, it is I....I am Wisdom. Mine is the blast of the thundered Word by which all things were made....I am Life."[5]

Another medieval document, in which the theory of the microcosm and macrocosm finds expression, is the *Book of Nature*, generally called by the name of Konrad von Megenberg, who translated it into German in the fourteenth century from a Latin text, compiled in the thirteenth century by a pupil of Albertus Magnus. The first section of the book begins with an account of the nature of humanity. We are told that God fashioned man

---

[1] Cornford, F. M. (1937), *Timaeus*, 41 D, p. 142.
[2] Cornford, F. M. (1937), pp. 38–9.
[3] Clarke, J., and Geikie, Sir A. (1910), p. 126.   [4] Singer, C. (1917).
[5] Modified from the translation in Singer, C. (1917), p. 33.

on the same lines as the world as a whole, and, as the sun is set in the midst of the planets, so man's heart is set in the midst of his body; from the likeness of man to the world, he is called in the Greek speech the microcosm or little world—thus, within one man's coat, the whole cosmos stands revealed.[1]

The idea of the spirit of the universe as something comparable with that of man, was expressed in more recent times by Kepler,[2] in whose writings medieval and modern views of the world jostle one another strangely.[3] Later in the seventeenth century, Leibniz held that each soul is so constructed as to represent the universe in its own manner.[4]

Mystical apprehension of the relation between man and the Whole, by no means exhausts the medieval conception of microcosm and macrocosm, which was applied to minutiae as well as to broad generalities. Hildegard, for instance, gives an account of the human body, comparing its organs to the constituent parts of the universe. Her work epitomizes the science of her time, which she fits with extreme ingenuity into the framework set by the theory.[5]

Though the over-elaborated belief in the microcosmic nature of man led to many absurdities, it sometimes opened the way to new and sound conclusions. It has been shown,[6] for example, that Harvey's discovery of the circulation of the blood (1615–19) was based, consciously, upon two Aristotelian tenets—that of the perfection of circular motion, and that of parallelism between the macrocosm and the microcosm. Harvey regarded the circulation of the blood as the reproduction, on microcosmic lines, of a pattern which is revealed also in the universe as a whole.

It is not surprising to find that Harvey's younger contemporary, Sir Thomas Browne, came to accept the doctrine of the macrocosm and microcosm, for it was exactly suited to his turn of mind—understanding it, as he did most things, in a sense of his own. He wrote in the *Religio Medici*: "to call ourselves

---

[1] Pfeiffer, F. (1861), pp. 3–4.
[2] Kepler, J. (1858–70), vol. v, 1864, *Harmonices Mundi*, 1619, p. 266.
[3] Cf. Singer, C. (1941), pp. 200 *et seq.*   [4] Carr, H. Wildon (1930), p. 162.
[5] Singer, C. (1917), pp. 30–5.
[6] Pagel, W. (1951); see especially p. 28.

a Microcosm, or little World, I thought it only a pleasant trope of Rhetorick, till my neer judgment and second thoughts told me there was a real truth therein. For first we are a rude mass, in the rank of creatures which onely are, and have a dull kind of being, not yet priviledged with life. . . ; next we live the life of Plants, the life of Animals, the life of Men, and at last the life of Spirits, running on in one mysterious nature those five kinds of existences, which comprehend the creatures, not onely of the World, but of the Universe."[1] Elsewhere he summarizes the doctrine in the aphorism, "There is no man alone, because every man is a Microcosm, and carries the whole World about him."[2] Sir Thomas Browne's ideas had, indeed, diverged widely from those of the *Timaeus* myth, for, to him, even the macrocosm in its greatness has become subordinate to man. As he wrote in one of his supreme passages: "whilst I study to find how I am a Microcosm, or little World, I find my self something more than the great. There is surely a piece of Divinity in us, something that was before the Elements, and owes no homage unto the Sun."[3]

In such utterances as these of Sir Thomas Browne, analogy is the handmaid of poetry rather than of the more rigid disciplines; but, as we have seen, it has its use also in science, in offering suggestions for hypotheses. It has a value, moreover, in the process of exposition; it may illuminate a problem, without actually serving as a technical instrument for solving it. Sometimes an illustrative analogy may transcend the reasoned conclusions of the man who offers it, and may thus constitute an advance which is none the less real because it may come, in part, from a region outside conscious awareness. The genius of a biologist may indeed show itself more clearly in his choice of the analogies under which he views his problems than in his orthodox scientific procedure. A case in point is Galton's symbolic description of 'particulate' inheritance,[4] which he put forward some little time before the rediscovery of Mendel's work, and which foreshadowed certain twentieth-century accounts. He wrote that "many of the modern buildings in Italy are

[1] Browne, Sir T. (1928–31), vol. I, *Religio Medici* (written 1635; 1682 ed.), pt. I, sect. 34, p. 43.   [2] Ibid., pt. II, sect. 10, p. 90.
[3] Ibid., pt. II, sect. 11, p. 91.   [4] Galton, F. (1889), pp. 7–9.

historically known to have been built out of the pillaged structures of older days. Here we may observe a column or a lintel serving the same purpose a second time.... I will pursue this rough simile just one step further, which is as much as it will bear. Suppose we were building a house with second-hand materials carted from a dealer's yard, we should often find considerable portions of the same old house to be still grouped together.... So in the process of transmission by inheritance, elements derived from the same ancestor are apt to appear in large groups."

Galton had a special penchant for analogies, and the passage just quoted was not the only example in which, by using them, he "builded better than he knew". Another instance is that of stable forms or groupings in inheritance, which he elucidates by comparison with the features of governments, crowds, cookery, etc., in all of which he finds that a limited number of characteristic groupings are recurrent.[1] Such expository use of analogy is a great help, provided we never forget that the analogies, which the human mind perceives, have an almost universal trend towards vitiating abstract relations by presenting them in terms of sense-relations; we cannot help craving for the relief from mental effort which is provided by 'picture thinking'. To a great extent proverbs are the outcome of this tendency. When we use such expressions as, "Strike while the iron is hot", or "The watched pot never boils", we are objectivizing and thus falsifying, a state of mind. Another, and more dangerous pitfall, is hidden in the very nature of analogies themselves. Every grade can be traced between remote analogies, and analogies which are so close that they pass into identities, and—paradoxically enough—it is often the remote analogies which have the greatest value,[2] while it is the close analogies of which we have to beware. The misuse of an analogy by pressing it to the point at which it is confused with an identity, is one to which biological thought is peculiarly liable. It beset the comparison between animals and plants, which was a sheet anchor of Greek biology[3]—a

---

[1] Galton, F. (1889), p. 22.   [2] Cf. Mill, J. S. (1843), vol. II, p. 429.
[3] Cf. Arber, A. (1950), pp. 11-14.

comparison which would have been helpful if it had been kept in its place, but which was carried so far that it led to serious misunderstanding of vegetable structure. Marcello Malpighi, in the seventeenth century, was one of those who placed too much reliance upon this resemblance. Having met with great difficulties in the anatomical study of the higher animals, he turned to plants, in the hope that these simpler creatures would reveal clues to animal anatomy. Though he failed to unravel his puzzles by this means, his results showed how valuable even a mistaken analogy may be, for, in pursuing it, he founded, incidentally, the science of plant anatomy. Nehemiah Grew, Malpighi's English contemporary, summed up the limitations of the plant-animal analogy in the words: "If any one shall require the Similitude to hold in every Thing; he would not have a *Plant* to resemble, but to be, an *Animal*."[1]

The failure to distinguish between an analogy and an identity has recurred again and again in the history of the evolution theory. One of the earlier writers on the subject, Lamarck, rightly perceiving a certain analogy between the individual and the race, concluded that the well-known effects of use, and of environmental factors, upon the structure of the individual, could be postulated also for the race.[2] His theory, however, broke down, because he did not realize that, though such effects may last for life in the individual, they are not passed on from generation to generation. In other words, the error in Lamarck's interpretation was due to the fact that the analogy between individual and race was less complete than he assumed it to be. Other examples of too facile an acceptance of this analogy are the idea that evolutionary phases are recapitulated in embryonic development, and the notion that there is a racial senescence comparable with individual senescence.

Darwin's theory of natural selection, again, depends largely upon an analogy—that between the controlled breeding of domestic animals and plants, and the whole historic develop-

[1] Grew, N. (1682), p. 173; on Grew and Malpighi, see Arber, A. (1941, *a, b, c*) and (1942).
[2] It is not easy to do justice to Lamarck's views in a brief summary; for a study of his work, see Nordenskiöld, N. E. (1950), pp. 316–30, etc.

ment of the organic world; one of the weaknesses of his theory lies in his failure to recognize the degree of incompleteness of this analogy. Another evolutionary thinker of a slightly later date, whose mind was unduly subject to the influence of similitudes, was Samuel Butler. His analysis and comparison of memory and heredity is a brilliant performance, but his theory is unsound because he mistakes the analogy between them for actual identity. The mechanistic theory of the organic world is rooted in the same misapprehension. The relation, for instance, between a living being and a machine, is typically the relation between a man and the contrivances which he has himself constructed, as it were in his own image, to reinforce and supplement his neuromuscular equipment. There is thus a natural parallelism between such machines and the organism of whose activity they form an extension. It is only in a strictly limited sense, however, that the idea of a machine, a derivative of the living creature, can be used in interpreting the characteristics of that creature. The analogy is obviously incomplete, but it may well provide demonstrations which bring the working of certain aspects of vital activity into a clearer light.

An analogy rejected as valueless because it is wrongly equated with an identity, sometimes contains a hidden truth, which may, later on, be seen for what it is. An example may be found in the resemblance between 'simple' and 'compound flowers'—a resemblance carried, not infrequently, into the minutest details of form and coloration. In the early days of botany, 'compound flowers' were regarded as being just 'flowers'. When increased knowledge revealed that the compound flower was an inflorescence, the idea of an analogy between the two types was discarded altogether, and for a time the striking resemblances between 'simple' and 'compound flowers' were neglected and ignored as meaningless. Recently, however, the analogy between them has again been brought into the light, and shown to have a significance of a much subtler kind than that at first attributed to it.[1]

---

[1] Troll, W. (1928); for an account of Troll's theory, see Arber, A. (1937), pp. 179–82, or (1950), pp. 144 *et seq.*

The present tentative sketch of the part played by analogies in the service of science, seems to show that, so long as their peculiar character is borne in mind, they are irreplaceable tools; but the essence of this character is *imperfection*. When it is forgotten that "no simile can be effective without an awareness of dissimilarity",[1] and when this forgetfulness leads to a handling of analogies as if they were complete and perfect, they grade into identities, and thus lose their *raison d'être*. It is their very imperfection which sets them in the boundary region of scientific thought, where they can exercise their unique power of acting as connecting links with other worlds of experience.

[1] Daiches, D. (1940), p. 54.

CHAPTER V

THE BIOLOGIST AND THE WRITTEN WORD

WE have now to consider the fifth stage of the biologist's labours—the process whereby he hands over to the common stock what he has seen and thought, so that it becomes part of the "comprehensive universe of discourse".[1] The arts can be conveyed from generation to generation, and from one nation to another, without the use of words, but, though biological understanding can, to a certain degree, be communicated through direct visual channels, it is kept alive and transmitted essentially by means of language. The biologist, in the more advanced stages of his work, is thus bound to take upon himself the function of a writer, whether he will or no.

In setting his findings into a permanent verbal mould, the research worker meets with various initial difficulties. One of these is a certain hesitation to submit his ideas to the testing ordeal of the pen, which often reveals unsuspected confusions of thought; the child who said, "How do I know what I think till I hear what I say?",[2] had got to the root of the matter. Many biologists must have had the experience of echoing, ruefully, Descartes's confession: "souvent les choses, qui m'ont semblé vrayes, lorsque i 'ay commencé à les concevoir, m'ont parû fausses, lorsque ie les voulu mettre sur le papier".[3]

Another hindrance, not peculiar to science, is that, by the limiting convention both of tongue and pen, words can be placed only in simple linear sequence, temporal in speech, but translated into spatial order in writing. The experience of one's own thinking suggests that it moves, actually, in a reticulum (possibly

[1] Adler, M. J. (1927), p. 200. [2] Blanshard, F. B. (1949), p. 76.
[3] [Descartes, R.] (1637), p. 66; (1947), p. 66. Translation, Haldane, E. S., and Ross, G. R. T. (1911–12), vol. I, p. 122.

45

of several dimensions), rather than along a single line. Even those who cannot accept the reticulum metaphor, might agree that thinking is like a river, which includes eddies and still backwaters, though, considered as a whole, it progresses in one direction. Neither a reticulum nor a river can be symbolized adequately in a linear succession of words. A written account is a mere thread, spun artificially into a chain-like form, whereas, in the weft of thought from which it is derived, the elements are interconnected according to a more complex mode. Haller recognized this nearly two hundred years ago, when, in speaking of relationships within the monocotyledons, he said: "Nature has linked her kinds into a net, not into a chain; men are in-capable of following anything but a chain, since they cannot express in words more than one thing at a time ".[1] The incapacity which Haller regretted, is occasionally overcome in non-scientific literature. Laurence Sterne, and certain modern writers influenced by him in their technique, have visualized, and tried to convey in language, the complicated, non-linear behaviour of the human mind, as it darts to and fro, disregarding the shackles of temporal sequence; but few biologists would dare to risk such experiments. Recognition of the non-linear character of thought in general, is exemplified in Bosanquet's[2] attempt to substitute 'implication' for the serial, chain-like mode of argument characteristic of conventional logic and 'discursive' reasoning.[3] With this attempt biologists may well feel some sympathy; it probably accords better with their views than with those of physicists and chemists.

A grave difficulty of another kind, which closely affects the biologist, is that his mode of exposition, if it is to be at all lucid, has to follow lines distinct from those on which his results have been reached in practice.[4] This is analogous to the difference, of which Aristotle was aware, between the very nature of a thing,

---

[1] Haller, A. von (1768), vol. II, p. 130: "Natura in reticulum sua genera connexit, non in catenam: homines non possint nisi catenam sequi, cum non plura simul sermone exponere."   [2] Bosanquet, B. (1911), (1920).
[3] Walsh, W. H. (1946), p. 62; on the nature of 'discursive' reasoning, see Chapter III of present book, p. 29, footnote 3.
[4] The difference between the process of discovery and that of demonstration is emphasized in Singer, C. (1941), pp. 227–8, etc.

and the way in which an origin for it may be represented. He writes, of his contemporaries at the Academy: "they say that in describing the genesis of the world they are doing as a geometer does in constructing a figure, not implying that the universe ever really came into existence, but for purposes of exposition facilitating understanding by exhibiting the object, like the figure, in process of formation".[1] The shape of a sphere, for example, is best understood, when it is described as arising by the rotation of a semicircle, even though it was not, in reality, actually produced by this method. In the same way the gesture is justified, when a man, asked what a spiral staircase is, explains it by a spiral movement of his hand, though no one imagines that it originated by means of any such motion. In biology, a case comparable with those just cited would be a botanist's description of a carpel as 'resulting' from the infolding of a leaf, or of a gamopetalous corolla as 'arising' from the fusion of free petals. Neither structure, in fact, 'became' in this way, but the character and relations of the ultimate product can be visualized better when this manner of expounding them is used.

The profound difference between the mode of discovery and that of exposition is far from being peculiar to biology; in many other fields of the intellectual life conclusions are reached intuitively, and it is only afterwards that a public highway to them is constructed. The matter is put in a nutshell in a saying attributed to Gauss, the great mathematician—"I have had my results for a long time; but I do not yet know how I am to arrive at them".[2] It appears, correspondingly, that the geometrical form, which Newton used in expounding the *Principia*, bears no resemblance to the mental processes by which his results came into being.[3] In philosophy, again, the artificiality of Spinoza's demonstration of his ideas, '*more Geometrico*', has been noticed repeatedly; it is impossible to suppose that it represents the route which he himself followed on the way to his conclusions. Similarly it is hard to imagine that Hegel did, in fact, reach his

[1] Cornford, F. M. (1937), p. 26, quoting Aristotle, *De caelo*, 279, b. 33.
[2] Quoted in Nelson, L. (1949), p. 89, and Beveridge, W. I. B. (1950), p. 145; the present writer has been unable to trace this dictum to its original source; cf. also Wright, G. H. von (1951), p. 22. [3] Keynes, J. M. (Lord) (1947), pp. 28–9.

conception of the Absolute by starting from Being, and pursuing the long dialectic path which he signposts for his readers.[1]

The biologist's need to 'put' his work differently, in order that it shall be understood by other minds, means that the whole research has to be reshaped in thought, before it can take written form. This is indeed a sheer necessity; if the worker tried to detail all the blind-alley routes which he essayed before finding his eventual path, the reader would be lost in the intricacies of a tortuous maze, and would give up the attempt to follow him. This need for a switch-over of method, from the work to its exposition, may be in part the explanation of the reluctance and distaste biologists often experience when the time comes to 'write up' their work, however much they enjoyed it in the doing; for the process of setting pen to paper is liable to seem alien and even disingenuous. There is no escape, however, from the compulsive truths that the biologist must write, and that writing is an art, and, moreover, a highly symbolic art. The biologist's picture of what he has observed, and his thoughts about it, can be imparted only by rows of little conventional marks on paper—a limited repertoire, the significance of which depends entirely on an agreed tradition. The verbal or pictorial formulation of scientific thinking is conditioned, as is any other kind of expression, by its chosen instrument. Whether a biologist is describing an organism in words, or drawing it, he has the same object before him, and in either case his intention is identical—that of delineating it in a way that will communicate his view of its character to other minds; but the portrayal in these two media is entirely different. Each is, on the one hand, limited, and, on the other hand, endowed with its own special significance, by the nature of the craft. The verbal medium in itself presents great inherent difficulties, but scientific workers have a way of turning a blind eye to these obstacles, and of failing to realize that to acquire command of writing demands a more exacting mental discipline than to become expert in the most refined laboratory technique. By many of us, mastery of

[1] Hegel, indeed, makes it clear that the final result of the process is also its logical ground, present in it all along; cf. McTaggart, J. McT. E. (1922), pp. 63–4, etc. On Hegel's dialectic, see also Chapter IX, p. 109–11, of the present book.

the verbal medium can be attained only by severe and prolonged effort; but no one who has to read scientific literature can doubt the worth-whileness of even the heaviest price paid by its authors as their dues for full initiation. Yet many biologists seem to look upon the process of writing as a tiresome mechanical task, to be slurred over with the minimum of exertion. The results are lamentable for the author as well as for the reader; chaotic language reacts upon thinking and reduces it, also, to chaos. That thought and its expression could not be separated was felt with special keenness by William Blake, though he was far from being a pedantic slave of language. "I have heard many People say", he wrote, "'Give me the Ideas. It is no matter what Words you put them into'." To this he replies, "Ideas cannot be Given but in their minutely Appropriate Words".[1] This dictum is as true for science as for the humanities, and it might well be inscribed on the walls of every biological laboratory.

When Plato made Socrates declare that any discourse ought to be a living creature with its members in due proportion,[2] he laid his finger on the crucial quality of scientific writing. Coleridge must have had this passage in mind when he alluded to "that *surview* which enables a man to foresee the whole of what he is to convey, and arrange the different parts according to their relative importance,. . .as an organized whole".[3] Imaginative writers sometimes feel that what they have penned has developed an independent vitality of its own. In Dante this feeling rises to white heat; he personifies his odes, addressing them as living creatures. Biological writers, at their best moments, may in a small way, on their humble plane, have an inkling of the delightful consciousness that what they have written has come alive under their hands—indeed, unless this happens, there is little chance of the result forming an organism rather than an artefact. It is a misfortune that the ultra-objective attitude, which scientific tradition encourages in the biologist, often militates against the vitalizing of his work into an organic whole. 'Facts' loom so large, and there is so much

[1] Blake, W., in Keynes, G. (1935), p. 101 (pp. 60–2 of the Rossetti note-book).
[2] Jowett, B. (1871), vol. 1, *Phaedrus*, 264, p. 599.
[3] Coleridge, S. T. (1817), vol. 11, p. 58.

justifiable fear of the personal element in interpretation, that what is offered is frequently a mere heap of juxtaposed data, rather than something characteristically alive, in which all the parts are related both to each other and to the whole. A record of research should not resemble a casual pile of quarried stone; it should seem "not built, but born", as Vasari said in praise of a building.[1] As in architecture, so likewise in creative scientific work, the expression should have a content extending beyond the writer's own direct vision. The architect of a cathedral, however vividly he may visualize the final outcome of his art, cannot possibly know beforehand exactly what impression the whole, or its individual parts, will convey from every standpoint, and under every future condition of light and atmosphere. It may well be that those who visit it in later years may detect aspects, exquisite in quality, which, though they are due to the architect's genius, never actually entered his awareness. Biological writing, also, may occasionally reach a level transcending the writer's consciousness, but this can happen only when the factual material is fused in the crucible of thought, a process demanding severe mental effort, not always congenial to the researcher, whose highly trained skill of eyes and fingers makes restless demands for employment, and tends to sap the energy needed for the more exacting activity of pure brain-work.

By what means the organization of the results of research is to be accomplished, is a knotty problem. After the main skeletal system has been, as it were, hewn out, there must be a further process of moulding subsidiary material into adequate relation with this skeleton, before the whole thing can come to life. The method which Thomas Hobbes followed in writing his *Leviathan* (1651) can be paralleled in the practice of biologists, who have to wrestle with their ideas in the crude mass, and to transform them into a significant whole. This method may have been suggested by the ways of his friend, Francis Bacon, who had liked in former years to dictate his thoughts to Hobbes in the "delicious walkes at Gorambery". Hobbes, we are told, first drew "the Designe of the Booke", and then "walked much and contemplated, and

---

[1] "non murato ma veramente nato." Quoted in Scott, G. (1914), p. 221.

he had in the head of his Staffe a pen and inke-horne, carried always a Note-book in his pocket, and as soon as a notion darted, he presently entred it into his Booke, or els he should perhaps have lost it.... Thus that booke was made."[1]

One of the factors inimical to organic wholeness in biological writing is the excessive use of quotation. An attempt, for instance, to set forth the present position of any controversial question, or to deal with a problem on historical lines, is apt to degenerate into a series of heterogeneous citations, on which the reader can scarcely fix his attention, and which give the impression of a lifeless compilation of detached notes. The extensive use of sheer quotation, though it seems at first sight a sign of modesty, is actually a self-protective device of the work-shy mind, which hankers to transfer mental labour from itself to its audience.

Hobbes not only realized, as we have seen, the essentialness of setting every thought in its due place in the organized whole, but he also saw that, before this could be done, each of these notions must be captured on the wing. Here the scientist has little to learn from others. The card-index habit, and the rule of immediately recording every observation or idea that possibly may be relevant, are commonplaces of scientific method; the notion, once caught and pinned down, can then be evaluated at leisure. More than a hundred years after Hobbes, Samuel Johnson, who showed other evidences of a latent capacity for scientific work, was most insistent on the importance of noting observations at the moment, and on the spot. He says, indeed, what a modern biologist might say, if he clothed his ideas in superb eighteenth-century language: "He who has not made the experiment, or who is not accustomed to require rigorous accuracy from himself, will scarcely believe how much a few hours take from certainty of knowledge, and distinctness of imagery; how the succession of objects will be broken, how separate parts will be confused, and how many particular features and discriminations will be compressed and conglobated into one gross and general idea."[2]

[1] Aubrey, J. (1950), pp. 150–1.   [2] Johnson, S. (1775), pp. 239–40.

The recognition of the value of immediacy is not the only feature in which scientific writing need not fear comparison with other forms of literature. A second no less admirable quality is that, at its best, it consists essentially of verbs and substantives, the sinews and bones of language. Few scientists produce work "overgarnished with Rhetorical Tropes, which like Flowers stuck in a Window...create a darkness in the place".[1] In a research worker's writing, simplicity and freedom from extraneous ornament arise naturally, because he has little temptation to write at all unless he has something to say; he is not just 'a writer', looking about for a subject on which to exercise his pen. This is fortunate, since, in Renan's words, "la règle fondamentale du style est d'avoir uniquement en vue la pensée que l'on veut inculquer, et par conséquent d'avoir une pensée".[2] Scientists sometimes evince a contempt for 'style', because they do not understand it in Renan's sense, but regard it as a sort of superficial 'literary' decoration, to be added in the final stages; but, in truth, style is the very essence of the work. If the thought is shaped with delicate economy, and has become strong, clear-cut and supple, and if the words materialize it with absolute precision, so that matter and manner are indissolubly fused, the elusive quality called style is won without wooing; but this recipe is as difficult to follow as it is easy to enunciate.

When he compares other intellectual disciplines with his own, the biologist may well feel that the literature of his subject excels in the stress laid upon the meticulous citation of authorities; even if this is sometimes carried to an absurd length, it is a fault in the right direction. In reading books on philosophy, for instance, he cannot but be struck by the way in which authors of the present day often make little attempt to distinguish what is original in their own writing from what is derivative. The result is that the student has to labour through much that is a mere reproduction—sometimes an inadequate one—of views expressed by others long ago, before he can sift out what is new. No

---

[1] Sharrock, R. (1672), p. 3. (This, which is the second edition, is quoted because, in the only copy of edition 1, 1660, available to the writer, the corresponding page is missing.)
[2] Renan, E. (1883), p. 220.

doubt the nature of the topics treated makes this, up to a point, inevitable, but the biologist, who has been brought up in the tradition of indicating exactly how much he owes to his predecessors, cannot but feel that a less high-handed method than that of the philosopher is more helpful to fellow-students.

Apart from those major hindrances in conveying his meaning, upon which we have already touched, the biologist has to negotiate certain obstacles of a minor kind. One of these lesser traps is that, being generally accustomed to express himself in lecture form, he is liable not to make adequate allowance for the fact that the written word is necessarily shorn of nuances, such as those conveyed by variations in pace, tone, and emphasis, which help so much towards the comprehension of speech. In writing, these nuances may be in part translated into terms of punctuation. Since lucidity of expression is the biologist's chief need, he often finds the 'heavier' system of stopping the more useful. In this system it may be said, roughly, that the stops are used to indicate such pauses as a good reader would make in trying to 'put across' the meaning as intelligibly as possible. In the written version of speech, however, the order of the words, and the internal and external balance of sentences, are even more serviceable than punctuation in supplying the place of renounced vocal subtleties. As Sir Kenelme Digby wrote in the mid-seventeenth century: "And such an effect as the manner of gesture and earnestnesse worketh in speaking, the like doth the manner of couching the sense, and the phrase, in writing."[1]

For the biologist yet another source of solicitude reveals itself—the difficulty inherent in the very nature of words. In the physico-chemical disciplines, and in those aspects of biology which are modelled on their lines, words tend to be treated as sharply differentiated and fixedly definable entities. Such discontinuity and limitedness accords ill with the quality of the living thing, which is not only always changing, but has a borderland or aura, shading off from the core of its own individuality into the environment.[2] There is no reason, however, why the

[1] Digby, Sir K. (1654), Dedicatory Letter To the Lady Digby.
[2] Smuts, J. C. (1926), p. 18.

53

biologist should not, for his own purposes, make more use of the literary aspects of language than the physicist finds necessary. Instead of thinking of his writing as a mosaic of separate words, each with its rigid and inescapable denotation, he may recognize that the *whole* should dominate the elements into which it may be analysed, and that individual words, if dissected out of this totality, lose their value through the mutilation due to their uprooting. He is at liberty, also, to use the enriching overtones of associative meaning, which can be called into play to evoke complex vistas of thought, and to "insinuate further meanings then the meer words barely considered, do seem to imploy".[1] The biologist's task is set in the debatable country between physical science and the humanities, and, while taking all he wants from the language of humane literature, he must also use sharply defined scientific terms. Unfortunately, when these are not new words, they are often adapted to this special use by discarding their atmospheric overtones, and also by paring away some of the fullness of meaning which they have acquired by organic growth in the course of age-long development. When, for instance, to form a new biological term, a word is adopted from common speech, its connotation is generally narrowed. Such ancient folk-names as 'leaf', 'seed', or 'root', each covers a much wider field in the mother tongue than when used in a strictly botanical sense; scientifically a grain of wheat ceases to be a 'seed', and a rose leaf is not (except in a certain hypothetical sense) a 'leaf'. The meaning of a word which becomes a scientific term may be reduced in depth, also, as well as in width; the word 'species', for example, is less significant to-day than it was in the pre-scientific phase, in which it related to the eternal 'Form' incarnated in each individual.[2]

The sequence of phases, through which scientific language passes historically, is fraught with meaning. Man's first observations of natural things cannot but have been couched in his primitive current vocabulary. Then, as the scientific outlook was gradually extended and defined, a technical terminology became

[1] Digby, Sir K. (1654). Dedicatory Letter To the Lady Digby.
[2] Cornford, F. M. (1932), p. 78.

a necessity, and thus the subject entered on a phase in which it ceased to be 'understanded of the people'; but this was merely transitional. When the language, in which science finds expression, is incomprehensible except to the initiated, it is a sign that the mode of thought is not yet at the maximum. By the time the subject reaches its high-tide mark, its largest generalizations transcend technical limits, and can again find expression in common speech: "The wheel is come full circle". We may perhaps detect a symbolic parallel to this rhythm in the stages of initiation described for Zen Buddhism.[1] Before a man tries to understand Zen, mountains are mountains to him, and waters are waters. Then, when he is gaining insight through intensive study, mountains cease to be mountains and waters to be waters; but at the last, when the aspirant has reached the goal and is at rest, he sees mountains again as mountains and waters as waters. In such a sequence as this, and also in the growth of scientific language, words gradually shed the cramping limitations of the phase succeeding the novitiate, and gain significantly in depth. Thus 'mountains' and 'waters' mean incomparably more at the end than at the start.

In assessing the qualities of biological literature, the paradox must be accepted—however reluctantly—that sometimes poor writing may serve certain scientific purposes more effectively than writing which is intrinsically better. The biologist does not write for his own countrymen alone, but, equally, for those whose native tongue is other than his. He thus wishes his work to be both comprehensible to the foreigner who reads his language, also to be translatable for those who are not acquainted with it. Generally speaking, one condition for good writing is the control of an extensive vocabulary, provided with a wealth of alternative words, fitted to express different shades of meaning; another condition is complete command of idiom. It is unfortunate, however, that both a rich vocabulary and an idiomatic style are serious obstacles to comprehension by a foreign reader. The student, for instance, who is beginning to cope with German scientific work, will often find it easier to start with an author,

[1] Suzuki, D. T. (1927), p. 12.

who writes in that tongue without belonging to a German-speaking country, and whose language is thus less copious and less idiomatic than that of a native. Again, when translation is in question, writing not of the first quality is easier to transmute into the speech of another race. Nevertheless, to lay much stress on being easily understood may be a fatal mistake, for the sort of clearness which helps towards easy translation, and which is indeed demanded, by general consent, by the lay public which reads scientific books, may, when viewed closely, prove to be a pseudo-simplicity, due to sacrifice of essentials. A sketch of a landscape in pen-and-ink outline may be much *clearer* than a corresponding water-colour, but, in the painting, mistiness, and effects of hue and light, which can only be hinted at in a black-and-white drawing, can be given their full significance. The water-colour thus comes nearer to full representational truth than the sharply definite ink outline. The translation of a landscape into black-and-white demands a method which recalls, where writing is concerned, an unsparing technique of pruning and rejection, which eases matters for the audience by focusing upon a few selected factors; this process is liable to degenerate into skeletal abstraction, in which reality, in its concreteness, fades out of sight.

Extreme lucidity may, moreover, defeat its own ends, not only by over-pruning, but also by eliminating too completely those rough places, which keep the reader awake, and stimulate him to make an effort at understanding. Especially in French scientific literature, smooth clarity of language is apt to give the student so little to brace himself against, that, after running his eye agreeably over the pages for some time, he may become suddenly and distressingly aware that his attention has been so far beguiled by the limpid flow of words, that he has failed to grasp the *meaning* of what he has read.

The risks arising from over-lucidity and over-simplification are most conspicuous in popularizations of science. The trend of the present day is all towards the attempt to express at least the main results of scientific work in a form which those who have never worked at the subject will imagine that they can understand. It may be recalled that this is an aim which Sir Isaac Newton set before himself in a certain connexion, but which he afterwards abandoned. He first intended to write Book III of the *Principia* (which demonstrates the frame of the System of the World) in such a manner that it might be read by many, but he finally decided to reduce this Book, in the mathematical way, into the form of propositions comprehensible only to those who had mastered the principles established in the first two Books.[1] He thus guarded against the insidious dangers of an easy exposition, which deceives the reader into fancying that he has reached, by a primrose path, conclusions which cannot in reality be approached except by the long and toilsome struggle of a prepared intelligence.

The biological writer is not only expected to offer clarity, regardless of what it costs, but also to meet the cognate demand for extreme brevity, especially in scientific journals. The obligation to be brief may, it is true, have great advantages; Coleridge, whose prose tended to diffuseness, often achieved concentrated expression in the marginal notes wherewith he decorated his own (and his friends') books, since here he had no choice but to submit to confinement of space.[2] In scientific writing, however, brevity is often harmful rather than helpful. The literary man may sometimes aim at producing a gem-like aphorism by ruthless lapidary work, and may rejoice to watch it grow "small by degrees and beautifully less"; but such *absolute* brevity is a dangerous technique for the biologist to adopt—for him, verbal parsimony can never have more than a *relative* value. Brevity may be the soul of wit, but it can never be the soul of science. In general it can be attained only by sacrificing many details, and many qualifications of statements, which ought to form part of the record, even if they are in themselves dull and boring. If all such elements are eliminated, what the author has to say is expressed by what might be called short-cuts across country, whereas the goal of his thought can often be appreciated more fully if it is reached by a longer route which gives the mind

---

[1] Newton, Sir I. (1803), vol. II, pp. 159–60.
[2] Muirhead, J. H. (1930), p. 258.

of the reader a chance of acclimatization on the way. The student of biology has reason to be grateful to such a man as Bütschli, 'Architect of Protozoology', who believed that it was better to write too much rather than too little, and who therefore produced his monumental work (1889) on so comprehensive a scale that it has been said by a recent authority in this field still to hold its place as an indispensable source book, despite its age.[1]

The question of brevity in scientific writing is linked with that of the size of books. It has been suggested[2] that the 'conventional' limitation as to the amount that can be included in any one work exercises a potent influence upon its content. It is true, though within tolerably wide limits, that books comply with the demand for a certain average size, convenient for standardized bookcases; but it is doubtful whether the factor of word-number, to which book-size is closely related, can be called merely 'conventional'. Ideally the author of a book, which is not purely a work of reference, should write as much as (and no more than) the reader can make his own, at a normal pace, within a reasonable period. The treatment of the author's material is thus conditioned by the anticipated reactions of the reader, which depend, in their turn, upon the physical limitations of the human eye, and, even more, upon the restricted capabilities of the human brain.

Scientific writing, even at its most factual and descriptive, always involves a certain element of interpretation, so that the biologist finds himself confronted constantly with the problem of what 'explanation' really means in scientific work. Almost as many answers have been given to this question as there are writers who have attended to the subject.[3] Physicists appear to regard the unveiling of material and efficient causes as being, in itself, 'explanation'. The physico-chemical aspects of science often, indeed, lend themselves to such an account, but biology, being less abstract and less conventionalized, remains inexplicable, if the word 'explanation' is used so narrowly. The

[1] Dobell, C. (1951).   [2] Collingwood, R. G. (1924), p. 10.
[3] As a mixed sample of the possible meanings attached to the term *scientific explanation*, see Westaway, F. W. (1919), p. 242; Campbell, N. R. (1920), p. 113; Stace, W. T. (1920), p. 65; Thompson, W. R. (1937), p. 125.

biologist may well feel more sympathy with Bosanquet's broader view, according to which, to explain a thing "is to think it in terms of the whole".[1] Since we believe that biology is *intelligible* (though not *explicable* in physico-chemical terms) what we require is a notion of explanation which has a greater richness of content than one which has been prearranged for the inanimate. As a personal opinion it may perhaps be hazarded that *the biological explanation of a phenomenon is the discovery of its own intrinsic place in a nexus of relations, extending indefinitely in all directions. To explain it is to see it simultaneously in its full individuality (as a whole in itself), and in its subordinate position (as one element in a larger whole).* On such a definition, classification is treated as one of the corollaries of explanation.[2]

In complete isolation a phenomenon is merely 'brute fact', which from its very nature, can never be grasped mentally, but, when 'explained', this raw datum is transmuted into an 'intelligible' entity—that is to say, into a relational form which is no longer alien to the mind.[3]

[1] Bosanquet, B. (1911), vol. II, p. 305.
[2] Bosanquet, B. (1911), vol. II, p. 198.
[3] Cf. on philosophical explanation, Joachim, H. H. (1948), pp. 37–8, 51, etc.

PART II

THE BASES OF
BIOLOGICAL THINKING

INTRODUCTION

IN the first part of this book we considered five of the stages into which the biologist's work may be analysed, tracing it from the selection of a problem for research, through the processes of acquiring and marshalling factual data; the formation and testing of hypotheses; and the communication of the results to fellow-workers. We have now reached the sixth and final stage, which we have called that of *contemplation*. It must, however, be emphasized that the separation of the first five phases from the ultimate phase is a matter of convenience, rather than an essential distinction, for meditation should play its part in the biologist's work from its beginning to its end. We turn now, however, to that particular type of contemplation which demands energy which is not set free until the detailed work is finished (so far as any scientific work can ever be described as finished). The biologist should then be in a position to go beyond his individual observations and conclusions, and to assess his general mode of thought, and the relations of his work in a broader context. In so doing, he has to penetrate into the significance of various conceptions, which, up to this point, he has taken at their face value; he may, for example, brace himself to ask what, precisely, he has been meaning by 'truth'. He must also undertake the heavy task of unearthing his basic assumptions and their consequences; bringing them into full daylight; and subjecting them to unprejudiced criticism. Moreover, recognizing how easy it is to slip from scientific balance into partisanship, he must focus upon the various antithetic ways of viewing biological reality, and satisfy himself that his ideas have not been distorted by an exclusive adherence to one or other of two partial conceptions, such as 'form' and 'function'. Furthermore, he may try to understand how far the many antitheses of this kind can, in fact, be synthesized, or whether they must be left as irreducibly opposed. Finally, he must envisage the relation of his own thinking to the more abstract procedure of logic and metaphysics, and consider what biological thought may be able to offer in return for the often unrecognized largesse which it has received, throughout the ages, from philosophy.

If, after such considerations, he finds that he has arrived at something possibly worth saying, he is faced with even more taxing problems concerning the written word, than those he met with when setting forth the straightforward scientific results of his research. Difficult, however, as it may prove, he cannot shirk the task of expression, for, though some monastic contemplatives are satisfied with a wholly interior life, no biologist is likely to find such a withdrawn existence adequate. The scientific tradition, of full sharing of thought and experience, is happily compulsive; as one of the research fraternity, the biologist is committed, not only *ad contemplandum*, but also *contemplata aliis tradendum*.[1]

---

[1] Cf. Aquinas, St Thomas, in Migne, J. P. (1845, etc.), vol. III (1841), p. 1377, *Summa Theol.* II/II, quest. 188, art. 7; translation in Aquinas, St Thomas (1911, etc.), II/II, p. 283. (The writer is indebted to Mrs E. A. Bullough for the source of this quotation.)

CHAPTER VI

BIOLOGY AND TRUTH

ALL that we have said hitherto has been informed by the obvious idea that the biologist's aim is to arrive at the truth. Truth is, indeed, the concept in which all science centres, as well as all philosophy. Thought itself has even been defined by a logician as "mental activity controlled by a single purpose, the attainment of truth".[1] The same writer holds that the adjectives 'true' and 'false' represent standards imposed by the thinker upon himself, since they correspond respectively to the imperatives, 'to be accepted', and 'to be rejected'.[2] The research worker may be prepared to accede to these statements—with the probable proviso that they seem to him to be platitudes—but he may well feel that they merely postpone his difficulties, since he is left in the dark as to the criteria which the thinker should use in determining what to accept and what to reject; and this is, after all, the crux of the matter. Before the biologist can feel satisfied with his own research procedure, he has still to face the insistent question, 'What is truth?', even if he can hope to arrive at nothing more conclusive than the hint of a suggestion of the possibility of an answer.

So much modern work is founded—though often in complete unconsciousness—upon Spinoza's scheme of things, that, in trying to tackle any difficult subject, it is well to make a start by recalling what he wrote about it. In following, from one of his works to another, his view of truth, we find that it underwent change and development, and that the sequence of its phases offers a framework which we can use. In his early essay, the *Short Treatise*, he says that "Truth is an affirmation (or a denial) made about a certain thing, which agrees with that same thing; and Falsity is an affirmation (or a denial) about a thing, which

---

[1] Johnson, W. E. (1921–4), pt. I, p. xvii.
[2] Johnson, W. E. (1921–4), pt. I, pp. 7–8.

does not agree with the thing itself".[1] Kant, again, regarded truth as the agreement of knowledge with the object.[2] This is, indeed, the common-sense meaning still attached in ordinary parlance to the phrase, 'telling the truth'; it is the conception generally known as the 'copy' or 'correspondence' theory of truth. Why it should be called a theory, is difficult to see; it would be more accurately described as a definition, or (as it is sometimes called) a doctrine; a modern writer has stated it as the 'congruence' between the meaning of a proposition and a factual situation.[3] The 'copy' doctrine is applied instinctively by the biologist to the records of his observations, in which he is dealing with what are known as 'contingent' truths. This term is used for truths of which the opposite is not inconceivable. Suppose, for instance, that a botanist, examining a plant tissue under the microscope, observes that its cells contain starch grains; this is a contingent truth, since it is conceivable that the tissue might have contained none.

It is clear that the biologist's first aim should be so to formulate his statements that they 'correspond', as far as is humanly possible, with the observed facts. Here the physico-chemical disciplines have an advantage, for, if a result can be expressed metrically, it is often possible to indicate the margin of error, and thus to mark the degree of accuracy that has been attained. In the biological sciences, the necessity for such precision was realized relatively late. It is recorded,[4] for instance, that Charles Darwin had no idea of the need for critical accuracy in instruments; he used rough scales and rules, and did his measurements of capacity with an apothecary's measuring glass, which was, in fact, badly graduated.

Even when the importance of numerical exactness is realized, biological observations are often not such as to lend themselves to an expressible margin of error. The degree of exactitude can, however, be indicated, up to a point, by an account of the kind

---

[1] Wolf, A. (1910), pt. II, chap. xv, p. 102.
[2] "Übereinstimmung der Erkenntniss mit dem Object Wahrheit ist." Kant, I. (1902–38), Bd. III, 1904, *Kritik der reinen Vernunft*, 2nd ed; 1787, Elementarlehre, Th. II, Abt. I, Buch. 2, Haupst. 2, p. 169; translation, Smith, N. Kemp (1933), p. 220.
[3] Wood, L. (1940), p. 239.   [4] Darwin, F. (1888), vol. I, pp. 147–8.

and amount of evidence on which any statement is based. For example, the best modern work on the anatomy of any plant organ makes it entirely clear how many series of sections have been examined; exactly where they were located in the member in question, with particulars as to its age and size; and what is the actual distance between the individual sections figured. In all such studies, the advent of the automatic microtome in the latter part of the nineteenth century raised the standard of descriptive work, since it rendered possible a degree of spatial precision which was outside the range of the old hand-section technique.

A difficulty of a more subtle kind than that of making accurate observations, arises out of the fact that the biologist's 'copy' or record is, in its actual nature, something very remote from the original. He has either to use the symbolism of words—remembering always that they are "wise mens counters", but "the mony of fooles"[1]—or else some form of visual representation; each of these methods involves the translation of his perception into another medium. Since the biologist depends primarily on words to convey his results to others, he needs the warning which Hobbes gave three hundred years ago: "a man that seeketh precise *truth*, had need to remember what every name he uses stands for; and to place it accordingly; or else he will find himselfe entangled in words, as a bird in lime-twiggs; the more he struggles, the more belimed".[1]

The 'correspondence' doctrine of truth seems at first sight to be a simple copy-book maxim, easy to apply; but its apparent simplicity is due to its deficiencies. One of these is that it emphasizes unduly the objective side of truth, thus suggesting a hard-and-fast separation between the object and the perceiving subject. The subjective aspect of truth cannot, however, be ignored with impunity. No truth is truth for any man until he has rethought it for himself; it has, indeed, been argued that truth is not truth at all, except in so far as it is the living experience of a mind.[2] The Chinese philosopher, Mencius—who

---

[1] Hobbes, T. (1651), pt. I, chap. IV, *Of Speech*, p. 15.
[2] Joachim, H. H. (1939), pp. 13–14; cf. also Bosanquet, B. (1920), p. 150, "truth...is reality as it makes itself known through particular minds in the form of ideas".

5-2

was almost exactly contemporary with the first great European botanist, Theophrastus—voiced the same idea when he said, "Having-it-in-self is called true".[1] A truth, even if derived originally from an external source, when rethought, can become organically one with the thinker; if, however, he merely appropriates it, without working it into the texture of his mind, it bears as little living relation to his thinking as the bizarre, alien bits and scraps, with which the caddice-worm bedecorates its case, bear to the creature itself. Since what is true for a man is thus inseparable from his whole personality, the 'truth', even of a relatively simple object, may seem entirely different to two different observers, even if they are endowed with equally acute senses and intelligence; what is truth for either of them depends upon his individual field of interest, and the channels in which his mental life naturally flows. A Japanese artist once said that, after concentrating on the painting of the bamboo for many, many years, there was still a technique for the rendering of the tips of the leaves that eluded him.[2] A drawing of his would have been inspired by a conception of truth diverging widely from that of the Western botanical draughtsman, who would have produced a business-like delineation, faithfully showing the recognized specific characters, but who would not have given any special consideration to the question of how the subtle, individual play and curvature of the delicate leaf-tips could best be generalized. The same thing is constantly brought home to any biologist who is studying the literature relating to a particular topic. Again and again, on hopefully reading a memoir on the subject which is in his mind, he fails to find just the facts he needs, because the author's field of interest does not coincide with his own.

In addition to its undue objectivity, another defect of the 'correspondence' doctrine is that it is not easily reconciled with the continual *flow* of things. Plato in the *Timaeus* indicated that, because the visible world is a changing image or likeness of an eternal model, there can never be a science of Nature, since such

---

[1] Richards, I. A. (1932), Appendix, p. 43.
[2] Coomaraswamy, A. K. (1934), p. 41.

a science would involve a final statement of exact truth about an ever-changing object.[1] It is improbable that any biologist would accept without reservation the Platonic opinion of science, but few would dispute that, in reaching it, Plato had seized the essential fact that any scientific *system* of explanation has a certain static finality, and hence must be imperfectly compatible with the unceasing flux of Nature. He is putting us on our guard against the dangers of system-making, and emphasizing that the attainment of truth is a process which has no end. The biologist, when trying to express his own vision of reality, has no choice but to represent development and change by means of static statements. This method does no harm, so long as he never forgets that it is merely a necessary convention. His actual procedure is to adopt a stance, which is truth for him for the moment, and then to pass through it to a further standpoint, without, however, renouncing the conscious possession of earlier positions.[2] His progress thus resembles that of an advancing army, which, moving forward from point to point, is concerned primarily with the standing ground most recently achieved, but also retains control of all positions previously gained. Progress of this kind can be illustrated only by analogies drawn from life. It is incompatible with the mechanical metaphor that sees each addition to scientific truth as an individual brick added to a permanent fabric. The development of scientific truth is, on the contrary, like that of an organism; it does not grow by accretion of ready-made parts, as a building does. In passing from phase to phase, it suffers transformation from within, like an animal proceeding from the embryonic stage to maturity. No conclusions in science can be immortal—they serve their season, and, if they survive at all, it is in the form of offspring theories, in which certain of their characters live again in a new guise.

We began this study of truth by adopting, for the moment, the doctrine which we have just discussed, that the validity of a statement can be tested by its 'correspondence' with empirical data; this is the opinion indicated in Spinoza's earliest work, but

[1] Cf. Cornford, F. M. (1937), pp. 23–9.
[2] Cf. de Ruggiero, G. (1921), p. 120.

in later writing[1] he passed beyond views of this type, and, though not rejecting such cogency as the correspondence theory can actually claim, he assayed truth afresh by the different method of following out all the *implications* of the statement in question.[2] This method of testing truth is akin to the spontaneous procedure of the biologist, when he estimates some hypothesis by subjecting the deductions which can be drawn from it to experimental trial.[3] He attempts, that is to say, to discover how satisfactorily his hypothesis can hold its place in the general network of thought; for any valid hypothesis should be able to withstand the trial of exposure to a context of relations within relations. As the context widens in ever-increasing circles, so the hypothetical thought establishes its place in worlds of truth which are more and more inclusive. This conception of the meaning of truth has become known as the 'coherence' doctrine. It has been studied and elaborated in more modern times especially by Joachim,[4] who turned to it after scrutinizing and rejecting the 'correspondence' notion. He shows that the fact of correspondence is not truth itself, but is "at most a symptom of truth";[5] and he passes to the 'coherence' theory, which he expresses as the assumption that "Truth in its essential nature is that systematic coherence which is the character of a significant whole".[6] This definition stands or falls with the idea of the Uniformity of Nature.[7]

It may be well, at this point, to try to get a little closer to the problem of what kind of truth the biologist envisages, and how it should be related to philosophical concepts. A difficulty that has to be faced at the outset is that the elusive word, truth, includes several gradients of meaning. Two of its principal connotations have been distinguished by Hegel[8] as *Richtigkeit* and *Wahrheit*. *Richtigkeit*, which may be translated as 'correctness',

---

[1] [Spinoza, B. de] 'B.D.S.' (1677), *Tractatus de Intellectus Emendatione*, pp. 375–6; for translation, see White, W. Hale, and Stirling, A. H. (1899), pp. 32–3.
[2] Roth, L. (1924), pp. 54, 55, etc.   [3] See Chapter III of the present book.
[4] Joachim, H. H. (1939), first published 1906; see also Bosanquet, B. (1911), vol. II, chap. IX, pp. 263–94.   [5] Joachim, H. H. (1939), p. 28.
[6] Joachim, H. H. (1939), p. 76; cf. also the 'consilience' of Hobhouse, L. T. (1896), p. 414; a clear account of this is given in Metz, R. (1938; reimpression, 1950), p. 161.
[7] On the Uniformity of Nature, see Chapter VII, pp. 82–8, of the present book.
[8] Wallace, W. (1874), Sect. 172, pp. 262–3; and Mure, G. R. G. (1940), p. 165.

is the contingent truth of the empirical sciences, to which the correspondence test can generally be applied. It is that towards which the biologist has to struggle in the first place, and, since factual accuracy is so hard to attain, he may well feel that, having achieved it, he is justified in resting upon his laurels. Nevertheless, the contemplative element in his mind will not let him remain satisfied, in the long run, with truth that is merely provisional; as soon as he passes beyond this, he gets a distant glimpse of Hegel's *Wahrheit*—that truth in which the thought content is in agreement with its own essential character. The difference between *Richtigkeit* and *Wahrheit* may perhaps be made clearer by an example. A sketch of a landscape may claim *Richtigkeit* if it fulfils the requirements of correctness, both in being an accurate 'copy' of the landscape, and also in conforming to certain orthodox principles of landscape painting; but it may, despite this, be wholly inadequate as a picture. If, on the other hand, the artist sets aside 'photographic' representation, and, while faithful to the intrinsic nature of paint as a medium, portrays his own direct reaction to the visual aspect of the landscape, he cannot but achieve truth as *Wahrheit*.

Broadly speaking, the natural sciences, in the strict sense, and primarily the physico-chemical disciplines, live in the dry light of *Richtigkeit*; philosophy, on the other hand, looks to the very different illumination of *Wahrheit*. This antithesis between scientific and philosophical thought is to some extent synthesized by 'natural philosophy', if we may reserve this term for the mode of thinking which relies on both kinds of searchlight in order to irradiate the borderland region connecting science with philosophy. That aspect of biology, which cannot be brought under the sway of physico-chemical methods, finds its appointed home in natural philosophy.[1]

While *Richtigkeit* is connected closely with the 'correspondence' notion of truth, *Wahrheit* bears a definite relation to the 'coherence' doctrine, when this is carried to its highest pitch. On this view, if the relations of a true idea are examined, they are found to extend in an ever-increasing network, until, with

[1] On biology as a branch of natural philosophy, see also pp. 76, 125, 126.

no break in the reticulation, they finally pass beyond our limited ken, and embrace the whole of reality. Since the Whole can never be reached by discursive[1] thought, "all instances of truth will be relations...all truth, except truth of the thing itself, is conditional", as Lord Herbert of Cherbury said in the seventeenth century.[2] Equivalent ideas find frequent reiteration in modern thought.[3] The 'coherence' doctrine, developed as far as reason is able to take it, is thus necessarily imperfect, and we are left with the knowledge that the fullest truth that science can hope to achieve is truth which is relative, but extends by implication[4] beyond its own limits.

It has often been held that the sharp distinction marked by Kant between the *Phenomenon* and the *Noumenon* represents an antithesis which is in its very nature unresolvable. The terms of this so-called antithesis can be transformed, however, into gradients in a series, if we regard the *Phenomenon* as being the *Thing-in-itself*, seen as in a glass darkly by the limited and fitful illumination of relative truth, while the *Noumenon* is the same *Ding-an-sich*, which can be fully revealed only by the light of that absolute truth which is beyond our finite compass.

The idea that scientific truth is inevitably partial and relative, brings us to the consideration of error and falsity. Error must not be confused with meaninglessness; as Hobbes wrote in 1651, "if a man should talk to me of a *round Quadrangle*...I should not say he were in Errour, but that his words were without meaning; that is to say, Absurd".[5] If then we set aside statements that are without significance, and consider the category of actual falsity, we find much reason to accept Spinoza's dictum that erroneousness is not a positive quality, but that it consists simply in privation of knowledge.[6] From this point of view, error is truth in an imperfect and incomplete form;[7] and there is thus no duality

---

[1] On the term *discursive*, see Chapter III, p. 29, footnote 3.
[2] Herbert, E. (Lord Herbert of Cherbury) (1937), p. 88; see also p. 24.
[3] Cf. Bradley, F. H. (1922), vol. II, p. 675.
[4] On *implication*, cf. Bosanquet, B., Letter to Vivante, 1922, in Muirhead, J. H. (1935), p. 270.
[5] Hobbes, T. (1651), pt. I, chap. v, *Of Reason and Science*, p. 19.
[6] [Spinoza, B. de] 'B.D.S.' (1677), *Ethices*, pt. II, prop. XXXV, pp. 72-3. "Falsitas consistit in cognitionis privatione"; translation in White, W. Hale, and Stirling, A. H. (1930), pp. 80-1.   [7] Cf. Bradley, F. H. (1930), pp. 169 *et seq.*

of truth and falsity to reconcile, since we are left with a graded series of 'truths', beginning with those so imperfect and partial that they are classed as errors, and passing upwards through every gradation of conditional truth. This conception of error accords well with certain phases of scientific thought. The connexion between relative truth and falsity is brought home to us when we meet with two incompatible hypotheses concerning the same matter, each of which is supported by an amount of evidence which seems sufficient to guarantee the probability of its truth. For example, the structure of some organ of a living creature may be explained on purely mechanical grounds, as arising through physical and chemical necessity, or it may be explained teleologically, as formed for the express purpose of accomplishing some definite function. It is obviously impossible to dismiss either of these hypotheses as baseless, and to say that the other expresses the full truth. The most rational opinion appears to be that each of them is true, but incompletely so. Though such relative truths seem to us now mutually exclusive, at a higher level of cognition they may eventually reveal themselves as divergent aspects of the same reality. They will then both be included in a synthetic truth, approaching nearer to absolute truth than either of them individually. Compared with this eventual truth, either of the partial truths, which meet in it, may be called an error; they are falsifications of reality, but only in so far as they are incomplete. Moreover, this synthetic truth itself remains only relative, and will pass at last into a series of still truer stages. The final question is whether it is possible to reach beyond all these relative truths and to attain the kind of truth which is the goal of the 'coherence' process. Spinoza, having, as we have seen, lived through phases in which the 'correspondence' and 'coherence' doctrines were compatible with his thought, finally transcended them both, and, in the work of his maturity—the *Ethic*—he reached the ultimate position that truth cannot be tested by either of these two criteria, but that it is its own standard—*veritas sui sit norma*;[1] William

---

[1] [Spinoza, B. de] 'B.D.S.' (1677). *Ethices*, pt. II, prop. XLIII, Schol., p. 80; translation in White, W. Hale, and Stirling, A. H. (1930), p. 90.

Blake was visited by a kindred thought when he said: "Truth can never be told so as to be understood, and not be believ'd."[1] Truth being its own norm, it follows that the true idea, when once attained, is, as Spinoza said, beyond the reach of doubt.[2] In this conception he was echoing Plotinus, who had declared, fourteen hundred years earlier, that "the veritable truth is consistent not with an alien Reality, but with itself; it affirms nothing other than itself; it *is*, and what it is, that it affirms. How then might such truth be 'tested'?...For there can be found no Reality truer than the truth."[3] Such absolute truth can be achieved only through direct apprehension, or, in philosophic language, 'immediately'; it is vouchsafed only to minds prepared to receive it by long effort in discursive thought. If the emotions also are finely tempered, such minds may acquire what Newman called the Illative Sense.[4] By this expression he indicates a certain subjective capacity for grasping truth— a capacity based, not on reasoning alone, but on the personality as a whole. It may perhaps be identified with that special form of perception which, according to Lord Herbert of Cherbury, has the power of greeting truth with a ready and direct response.[5] Newman compares the illative sense to taste in art, which, however inborn it seems, owes much to accumulated knowledge, so completely digested, and so fully fused with emotional elements, that the certainty of discrimination, which it gives, has the appearance of being instinctive. When the illative power is lacking, one may easily persuade oneself that an idea bears the stamp of truth, when it does not. A flash of apparent intuition may be merely a pathological phosphorescence, simulating an absolute insight; but this false idea masquerading as truth should reveal its inadequacy when sanely and dispassionately scrutinized. A genuine and well-based intuition, on the other

---

[1] Blake, W., in Keynes, G. (1925), vol. I, p. 186.
[2] [Spinoza, B. de] 'B.D.S.' (1677), *Ethices*, pt. II, heading prop. XLIII, p. 79: "Qui veram habet ideam, simul scit se veram habere ideam, nec de rei veritate potest dubitare."
[3] Plotinus, *Enn.* v, v, 1, 2, in Dodds, E. R. (1923), p. 50.
[4] Newman, J. H. (1870); chap. IX, pp. 346 *et seq.*; cf. also Metz, R. (1938; reimpression 1950), p. 190.
[5] Herbert, E. (Lord Herbert of Cherbury) (1937), p. 148.

hand, will emerge unscathed from the most rigorous criticism which discursive thinking can apply to it.

It cannot be too much emphasized that, whenever the mind uncurtains something lying beyond relative truth, it is not by means of any magical sleight of hand; it is, on the contrary, the end result of the severest wrestlings of rational thought, stimulated yet controlled by disciplined emotion. We see this clearly in Plotinus and Spinoza, whose illumination was associated with intellectual effort so intense and so vitally irradiated by feeling, that in its incandescence it afforded them a distant glimpse of truth as absolute.

CHAPTER VII

THE BASIC ASSUMPTIONS OF BIOLOGY

WHEN the biologist has finished some specific task, and sets himself to see his results in a larger context, he may be compared to an artist of a reflective turn of mind, who, finally laying aside his palette and brushes, stands back from his easel, considers his picture, and focuses his inner gaze upon those initial agreed conventions without which his work could not have come into being. Hitherto he has taken these conventions for granted, and has concerned himself merely about whether his picture is good *of its kind*. Now that the time for alteration and retouching is over, he may feel that, *within the limits of its kind*, this is the best he can do; but, in this moment of thoughtful detachment, the question whether *the kind itself* will endure a critical scrutiny possibly arises in his mind. He may, for instance turn his thoughts to the evaluation of that symbolism by means of which he has translated his impression of a three-dimensional landscape into a two-dimensional form, or he may speculate on his own purely artificial allocation of a rectangular outline to his work. It may be objected that such considerations belong to the province of the art critic rather than to that of the artist, and that, correspondingly, the biologist, who anatomizes his own work in order to lay bare its basic assumptions, is leaving his own province and intruding into that of the philosopher. It is indeed true that the painter may not become a better artist through donning the critic's mantle; the unearthing of basic principles is, on the other hand, more closely cognate with the daily work of the biologist than of the painter. Though intellectual labours of this type may fall outside the narrower definitions of natural science, they find their place in the broader and more humanistic world of natural philosophy, into which, in the last resort, biological science is resolved.[1]

[1] See pp. 71, 125–6, of the present book.

No one can dispute that what suppositions are chosen for acceptance is of the utmost importance in deductive reasoning, since the stages of such an argument represent merely the unravelling of the significance of those principles with which the thinker starts; it may, indeed, be held that all systems of philosophy are methods of unfolding the implications of their own postulates. In the words that Plato puts into the mouth of Socrates, "every man should expend his chief thought and attention on the consideration of his first principles:—are they or are they not rightly laid down? and when he has sifted them all the rest will follow".[1] Men of science, on the other hand, may say that these considerations do not apply to them, since they are prepared to accept Aristotle's dictum that our knowledge goes back to premisses which are incapable of demonstration.[2] Dante was accepting this edict when he wrote in the *Convivio*, "no Science demonstrates its own subject, but presupposes it".[3] It is true, however, in science as well as in philosophy, that to refrain from criticizing first principles is often fatal. Obvious examples of the effect of a false basic assumption are to be found in such old paradoxes as that of Achilles and the tortoise. If it is supposed that duration is divisible into units, Achilles never *can* pass the tortoise, but since, in fact, he *does*, the fault evidently lies in this primary supposition.[4] In biology many apparently stable superstructures have collapsed sooner or later, because the assumptions, on which they were founded, proved to be unsound. It is not until the student of living things has satisfied himself as to the solidity of his foundations, that he will find a foothold—the ποῦ στῶ that Archimedes demanded as a prerequisite for putting his forces into action.[5] No foothold,

---

[1] Jowett, B. (1871), vol. I, *Cratylus*, 436, p. 711; see also vol. II, *Republic*, vi, 511, pp. 346–7, and vii, 533, p. 369.

[2] Mure, G. R. G. (1926), *Analytica posteriora*, bk. I, 3, 72 *b*.

[3] Dante in Giuliani, G. (1874), p. 142, Tract. II, cap. XIV, "perocchè nulla scienza dimostra lo proprio suggetto, ma presuppone quello". Translation, in Jackson, W. W. (1909), pp. 105–6.

[4] Cf., for instance, Pollock, Sir F. (1899), pp. 169 *et seq.* and Burnet, J. (1914), p. 84.

[5] "Give me a place to stand on and I move the earth", δός μοι ποῦ στῶ καὶ κινῶ τὴν γῆν, Archimedes, quoted in Pappus, viii; see Hultsch, F. (1876–8), vol. III, tom. I, p. 1060, 1–4.

however, can be more than temporary, for scientific work is in its essence fluid and progressive, and from time to time it outdistances the principles which formerly it had assumed as basic. It is thus necessary to scrutinize them repeatedly, and, if need be, to subject them to drastic revision; but it is a matter of great difficulty to overcome the natural disinclination to allow one's foundations to be probed and disturbed. As Kant wrote in the eighteenth century, "It is, indeed, the common fate of human reason to complete its speculative structures as speedily as may be, and only afterwards to enquire whether the foundations are reliable. All sorts of excuses will then be appealed to, in order to...enable us to dispense altogether with so late and so dangerous an enquiry."[1] Though this is true in general, to some adventurous minds the excitement of possibly shattering their scheme of things to bits, adds an attractive spice of risk to the search for flaws in those first principles which they have been accepting as incontrovertible. Kant had himself the courage necessary for uprooting his own convictions and inspecting them dispassionately; after he was awakened from his 'dogmatic slumber', and was led to criticize his basic principles, his researches in the field of speculative philosophy took a new direction.[2]

Basic assumptions may be regarded in two ways. They may be thought of *metaphysically*, in which case it is taken for granted that they are universally true, or they may be thought of *methodologically*.[3] This second alternative means that the biologist does not commit himself to their truth, but merely uses them as

---

[1] Kant, I. (1902–38), Bd. III, 1904, *Kritik der reinen Vernunft*, 2nd ed. 1787. *Einleitung*, III, p. 32; "Es ist aber ein gewöhnliches Schicksal der menschlichen Vernunft in der Speculation, ihr Gebäude so früh wie möglich fertig zu machen und hintennach allererst zu untersuchen, ob auch der Grund dazu gut gelegt sei. Alsdann aber werden allerlei Beschönigungen herbeigesucht, um uns...eine solche späte und gefährliche Prüfung lieber gar abzuweisen." Translation, Smith, N. Kemp (1933), p. 47.

[2] Kant, I. (1783), p. 13. *Prolegomena zu einer jeden künstigen Metaphysik*: "die Erinnerung des David Hume war eben dasjenige, was mir vor vielen Jahren zuerst den dogmatischen Schlummer unterbrach, und meinen Untersuchungen im Felde der speculativen Philosophie eine ganz andre Richtung gab". Elsewhere, however, Kant accounts differently for his awakening; see Caird, E. (1889), vol. I, p. 162.

[3] Cf. Woodger, J. H. (1929), p. 28; for a somewhat different approach see Stace, W. T. (1932), pp. 405–6.

convenient devices for purposes of research; the procedure is then to study what happens when they are treated *as if* they were true. If deductions from these assumptions cohere, and offer a rational picture, the resulting glow of satisfaction sometimes dazzles the biologist into forgetting that these deductions are valid only if the initial assumptions are themselves true. For example, he may choose to make the methodological assumption that, among living creatures, a group of species, if they closely resemble one another, must have a common ancestry. On this basis he may draw up a phylogenetic scheme showing the probable interconnexions and relationships of the individual species, and this, within its limits, may provide a useful working plan. Nevertheless this plan has absolute value only if the basic assumption as to the monophylesis of the group represents actual historic truth. If, however, the similarities between the species in question are due in reality to parallelism and convergence of lines of descent from different sources, the basic assumption is demolished, and the phylogenetic scheme, erected on the strength of it, falls to pieces.

Each basic conception underlying the reasoning of a working biologist, may be compared to one of his pieces of apparatus, such as a microscope. The average biologist takes his microscope for granted and uses it as a *tool*. His interest is concentrated, not on it, but on what he sees with it. A maker of optical apparatus, on the other hand, would cast a connoisseur's eye upon the microscope itself; he would criticize and appraise it as an *instrument*, without concerning himself especially about the observations which the biologist happened to have achieved with it. Passing from laboratory to study, the biologist may base the theoretical side of his work upon some accepted assumption, which he himself takes for granted as he does his microscope, without considering it his duty to search for evidence which might support or rebuff it. A philosopher, on the other hand, would be interested primarily in the biologist's underlying assumption. He would refuse to take it as 'given', but would examine it, and try to assess its validity, in and for itself, irrespective of the use to which the biologist was accustomed to put it.

The biologist, if asked *why* he narrows down his procedure by taking both his microscope and the fundamentals of his thinking for granted, might reply that life is so short that, if he attempted to become an expert in scientific instruments, and also a trained philosopher, no time would be left in which to pursue his own *métier*, in which neither the scientific instrument maker nor the philosopher could replace him. There is obviously a good deal in this argument, but a compromise is possible. Some technical knowledge of the microscope, and some scrutiny of the philosophic bases of his conceptions, might prove helpful to the biologist, even within the limits of his own work, though he could not hope to reach the *expertise* of the specialist in microscopes, or of the professional philosopher. The philosopher, as we have already noticed, claims, and with some justice, that it is one of his functions to criticize the basic notions which science takes for granted.[1] The biologist, however, may shrink from handing over his primary ideas unreservedly for the philosopher's adjudication, because the true inwardness of these ideas, in relation to the study of life, is likely to elude anyone who is not an initiate in that particular scientific field. Moreover it must not be forgotten that each philosopher starts with certain first postulates of the non-proven kind, and that these postulates sometimes strike the scientist as being singularly arbitrary. After all, though every one of the innumerable schools of philosophic and scientific thought is engaged in seeking the truth, the riddle of the universe remains unsolved; and, so long as this is so, it ill beseems any one group of thinkers to dictate *de haut en bas* to the rest.

In the mid-nineteenth century, John Stuart Mill held that the whole problem of the investigation of nature was comprehended essentially in the question, "What are the fewest assumptions which being granted, the order of nature as it exists would be the result?"[2] This pronouncement strikes the biologist of the present day as unduly dogmatic, but it has a real meaning from the point of view of the methodology of science. The reduction of basic assumptions to a minimum is indeed a help towards progress, since, being unanalysable 'givens', they oppose a blank

[1] Stace, W. T. (1920), p. 3.　　　　[2] Mill, J. S. (1843), vol. I, p. 560.

wall to scientific investigation. In a world offering data of this impervious type alone, science could not carry on its work. As a single example, from biology, of something which was long held, without adequate reason, to come into the category of the 'given', and to be thus immune from inquiry, we may recall the affirmation that the leaf is a basic unit of the plant body in the angiosperms. So long as this was assumed, any effort to understand the morphology of the leaf was forbidden; the leaf was a concept which one could not, as it were, get behind. When, however, the ban was lifted, and the leaf lost the privileged position accorded to it as an organ *sui generis*, the way was opened towards interpreting it. It might then be regarded, for instance, as an incomplete form of shoot,[1] an interpretation which—whether or not it is accepted ultimately—is at least an attempt to render the leaf-concept intelligible by placing it in a nexus of relations, instead of leaving it as an isolated fixed datum.

When we reflect, in general, on the various assumptions which are commonly regarded as basic for natural science, we see that the most fundamental of them is that the universe is intelligible to reason;[2] this is equivalent to saying that the universe is itself entirely rational, and penetrable to the intellect.[3] Such an assumption obviously involves a certain belief in the validity of the sense-impressions, and of the normal working of the human mind;[4] unless indeed, this belief is held, not only scientific thought, but all intellectual work of any kind, comes to a standstill. Despite the general consensus of opinion that Nature is intelligible, there exists also a recurrent tendency to doubt whether the universe can in fact be rationalized *completely*, or whether, in the last resort, an irreducible element of brute fact, or 'necessity', is left—an element which is altogether refractory to reason.[5] Plato evidently felt a degree of cogency in the latter

---

[1] On this theory, initiated by Casimir de Candolle, see Arber, A. (1950), chap. VI, pp. 70 *et seq.*   [2] Muirhead, J. H. (1931, reimpression, 1939), p. 231.
[3] Meyerson, É. (1931), vol. I, pp. 176 *et passim*.
[4] Thompson, W. R. (1937), p. 26.
[5] For an interesting consideration of the rationality or irrationality of the universe, see McTaggart, J. McT. E. (1922), pp. 169 *et seq.*, or 1st ed. (1896), pp. 172 *et seq.*

view, for he seems to have considered that there is something permanently chaotic within the cosmos.[1] Moreover, it is conceivable that certain features of modern physics hint at a basic irrationality underlying statistical uniformity. With this idea, however, the biologist is not concerned; within his particular world—which in scale is neither infinitesimal nor astronomically enormous—the principle of the rationality of Nature can be used, as an instrument of method, for working purposes.

When we come to inquire further into the meaning of intelligibility, we discover that it represents the mental realization of unity permeating the world of multiplicity. In other words, if intelligibility be accepted as a character of the universe, this signifies a recognition of the *Unity* or *Uniformity of Nature*. Of this Uniformity, the four primary 'Laws of Thought' have been regarded as the formal or mental aspect.[2] These Laws may be epitomized, albeit crudely, as: the Law of Identity (A is A); the Law of Contradiction (A is not both A and not-A);[3] the Law of Excluded Middle (A is either B or not-B); and the Law of Sufficient Reason (every consequent has a ground from which it necessarily follows). In the nineteenth century there was a tendency to reduce the principle of Uniformity simply to an inadequate version of the Law of Ground and Consequent, and to represent this Uniformity as limited to an inviolable sequence of cause and effect on the temporal plane. Such a conception is, however, unsound, since it assumes the possibility of analysing the multitude of phenomena into causes and effects arranged in chains—more or less parallel, though frequently entangled. It is supposed that these chains are isolable in thought, and that they can be imaged as consisting of discrete but firmly soldered links, each of which is the result of its predecessor, and bears a causal relation to its successor. This picture is strangely remote from the universe as we know it. Such imagery merely represents an effort at 'stringing out', in terms of time, that reality which

---

[1] Cf. Cornford, F. M. (1937), p. 361, etc.

[2] Cf. Bosanquet, B. (1911), vol. II, p. 216; for a general discussion of the Laws of Thought, see also pp. 209–16.

[3] The Laws of Contradiction and Excluded Middle are, obviously oversimplifications; see chap. IX, pp. 109–10 of the present book.

cannot be seen as it is except *sub specie aeternatitis*.[1] The word 'cause' is itself an unfortunate one; it apparently originated as a legal term,[2] and its scientific use tends to be distorted by associations which still cling to it from its forensic source. Although it is less than justifiable to equate the idea of Cause and Effect with that of the Uniformity of Nature, causation has a primary place in the methodology of science; exactly how it should be ranked, we will consider later in this chapter.[3]

A profound conviction of the unity and uniformity of Nature has been expressed by many thinkers at different periods. Beatrice tells Dante of the mutual order which all things observe,[4] and this conception was echoed, centuries later, by those who have held, as Leibniz did, that "l'Univers, quel qu'il puisse être, est tout d'une pièce, comme un Océan".[5] Newton, again, accepted the same idea when he wrote that "Nature is very constant and conformable to her self".[6] In the nineteenth century it was received as an unquestioned certainty. W. K. Clifford, for instance, said in 1872 that the instrument of scientific thought is "an observed uniformity in the course of events". By the use of this instrument, thought "gives us information transcending our experience, it enables us to infer things we have not seen from things that we have seen; and the evidence for the truth of that information depends on our supposing that the uniformity holds good beyond our experience".[7] Whereas Clifford thus makes the validity of the principle of uniformity depend upon induction from observations, John Stuart Mill, who believed that "every induction may be thrown into the form of a syllogism by supplying a major premiss", concluded that the uniformity of the course of nature is "the ultimate major premiss of all inductions".[8] Thus, while

---

[1] Cf. the discussion of Spinoza's idea of cause in Pollock, Sir F. (1899), pp. 149–51.
[2] Kneale, W. (1949), p. 61.   [3] See pp. 87–8.
[4]   "le cose tutte e quante
                        Hann'ordine tra loro;"
Paradiso, canto I, 103, 4. Dante Alighieri (1900), p. 362.
[5] Leibniz, G. W. (1875–90), vol. VI, 1885, *Essais de Théodicée*, p. 107, sect. 9 (first published, 1710).
[6] Newton, Sir I. (1931) (reprint from 4th ed. 1730), *Opticks*, bk. III, quest. 31, p. 376.
[7] Clifford, W. K. (1879), pp. 131–2.   [8] Mill, J. S. (1843), vol. I, p. 372.

Mill regards the principle of uniformity as the *basis* of induction, Clifford regards it as a *conclusion* from the same process; this incompatibility has continued to find expression in scientific thought, even to the present day.

If we look back to a time two hundred years before Clifford's pronouncement, we find Spinoza writing that Nature's "laws and rules, according to which all things are made and are changed from form to form, are everywhere and always the same".[1] We can appreciate how it was that Spinoza came to this view, if we recall that he founds his maturest work, the *Ethic*, on the conception of an absolutely perfect Being,[2] of whose Attributes two alone are known to men. These are Extension and Thought—that is to say, on the one hand, the familiar world of things, extended in space, and, on the other hand, the mental world. It is on the foundation of this monistic view that Spinoza develops his whole argument, and it follows inevitably that he treats the uniformity of nature as 'given'.[3]

In modern times a conception akin to that of Spinoza has been voiced by Charles Singer, who, writing as an historian, describes science as engaged upon a constant search for law and order in the universe.[4] Metaphysicians, also, sometimes speak of their own discipline in closely similar terms; it has been said, for instance, that philosophy "seeks to view the entire universe in the light of the fewest possible general principles, in the light, if possible, of a single ultimate principle".[5] Such interpretations of science and philosophy imply a confidence that a state of law and order exists, but that it is an unanalysable datum, which man's thought cannot penetrate, but may only reveal and make explicit. This belief is essentially intuitive; it recalls Einstein's opinion that the possibility that the regulations valid for the

[1] [Spinoza, B. de] 'B.D.S.' (1677), *Ethices*, pars III, p. 94, "naturae leges, et regulae, secundùm quas omnia fiunt, et ex unis formis in alias mutantur, sunt ubique, et semper eaedem". The translation is corrected from White, W. Hale, and Stirling, A. H. (1930), p. 105.

[2] It is pointed out in Joachim, H. H. (1901), p. 116, that though Spinoza seems to start with definitions of simple elements, and to construct the whole out of them, he is in reality arguing analytically from the whole.

[3] For an opposed view, which Spinoza also expressed, see pp. 85, 86.

[4] Singer, C. (1941), p. 2.

[5] Stace, W. T. (1920), p. 3.

world of existence are rational, is a faith springing from the sphere of religion.[1]

Even from the few dicta which we have cited, it becomes apparent that the status of the premiss regarding the uniformity of nature is singularly uncertain. It cannot be called self-evident, or a necessary truth; neither can it be proved inductively, because nothing short of a complete enumeration and study of every process in the universe, past, present, and future, could give it final verification. It is true that there is a large amount of evidence which appears to support it, but it is possible to put another interpretation upon this evidence. It is remarkable that Spinoza, in spite of having, in his final work,[2] enunciated and accepted the idea of uniformity, elsewhere in the same treatise took a different view, and said that the order, which we, in our ignorance, think we discern in Nature, is not actually an element in Nature herself, but resides in our 'imagination'.[3] In the next century Kant expressed a related view when he said, "The order and regularity in the appearances, which we entitle *nature*, we ourselves introduce. We could never find them in appearances, had not we ourselves, or the nature of our mind, originally set them there".[4] In interpreting this contention, it must be remembered that man is, in fact, an integral part of Nature—not an outsider and an alien—and, if order is implicit in the structure and functioning of his mind (using this term in a broad sense to include percepts as well as concepts) it is probable that order is implicit in the Whole of which he forms a part. Long ago, Lord Herbert of Cherbury denied the antithesis of Man and Nature, and his treatise, *De Veritate* (first published in 1624) has for its central conception "the organic relation between the impulse of Nature and

[1] Einstein, A. (1940), p. 605.
[2] See p. 84.
[3] [Spinoza, B. de] 'B.D.S.' (1677), *Ethices*, pars I, Appendix, pp. 37–8; for translation, see White, W. Hale, and Stirling, A. H. (1930), pp. 43–4.
[4] Kehrbach, K. (1919), Kant, I., *Kritik der reinen Vernunft*. Text der Ausgabe von 1781, Elementarlehre, Th. II, Abt. I, Buch I, Hauptstück II, p. 134. "Die Ordnung und Regelmässigkeit also an den Erscheinungen, die wir *Natur* nennen, bringen wir selbst hinein, und würden sie auch nicht darin finden können, hätten wir sie nicht, oder die Natur unseres Gemüths ursprünglich hineingelegt." For translation, see Smith, N. Kemp (1933), p. 147.

human thought ".[1] Notwithstanding his reservations, in practice Spinoza based his work upon the same belief, for he held that he could approach the truth about the Whole by the path of ordered argument, and he maintained that men's actions and appetites could be discussed as if they were as undoubtedly subject to rational law as lines, planes, and bodies.[2] The fact that he did, nevertheless, consider the possibility that Nature may include an element of irrationality, is perhaps merely a sign that he knew that the Principle of Uniformity could not be treated as proven. As we have seen, it is not open to any final inductive proof; and it is also evident that it cannot be proved deductively, since there is no more general principle from which it can be derived.[3] If, on the other hand, it is treated as a 'given' mystical intuition, it is seen as an assumption the basis of which falls outside the biologist's field of criticism. From the scientific standpoint, another approach is, however, possible; it may be suggested that the statement that Nature is uniform should be treated neither as a datum, nor as a conclusion from an argument, but as an *hypothesis*. That it should have its basis in intuition would then be natural; we found, in discussing the sources of discovery, that hypotheses often take their rise through processes including but transcending discursive thought.[4] We have already seen that hypotheses are not susceptible of actual proof; they advance merely to higher and higher degrees of probability as they are found to 'work'. This is what happens when we regard the Unity or Uniformity of Nature as an hypothesis; it is then visualized as being tested without intermission in innumerable pieces of scientific observation and experiment. It may, indeed, be permissible to think of the work of scientific research, in its totality, as a team-work attempt to discover whether experience altogether can be 'saved' on the hypothesis that there is a unitary Whole; or, in other words, whether the conception of the One, as opposed to the Many, is

---

[1] Carré, M. H., in Herbert, E. (Lord Herbert of Cherbury) (1937), p. 66.

[2] [Spinoza, B. de] 'B.D.S.' (1677), *Ethices*, pars III, p. 94, "humanas actiones, atque appetitus considerabo perinde, ac si Quaestio de lineis, planis, aut de corporibus esset". Translation, White, W. Hale, and Stirling, A. H. (1930), p. 105.

[3] Stace, W. T. (1932), pp. 349–50.   [4] Cf. Chapter II, pp. 20–1.

justifiable. On this view, the conviction that Nature is Uniform comes to the individual worker as one of the ultimate rewards of the scientific life, instead of being a facile preconception. A danger that must be faced, however, is that a scientist is often so deeply committed, from the start, to the idea of order in the universe, that it is difficult for him to test the hypothesis of the Uniformity of Nature with any impartiality. Surreptitiously, but unconsciously, he may have inserted the element of order into the basic formulation of his problem; when this has happened, it is not surprising if, at the end, he triumphantly finds it there.

Earlier in this book,[1] it was suggested that the biological *explanation* of a phenomenon involves the discovery of its own intrinsic place in a texture of relations. Such a definition assumes that phenomena in general are not disposed casually in an unorganized chaos, but in a nexus, which is an 'ordered whole'[2] of experience; this definition must thus rank as a corollary to the hypothesis of the Uniformity of Nature.

We have noticed[3] that the theory of causality is so closely related to the principle that Nature is intelligible and uniform, that it has sometimes even been identified with this idea. The various meanings that have been attributed to the causal principle recall the differing interpretations of Uniformity. Whereas Hume[4] held that we can detect only succession, and that the belief in a causal relation springs merely from the observation of customary sequences, Kant,[5] on the contrary, regarded this belief as an internal necessity of thought. This difference corresponds with that between those who consider that the results of observation do not preclude the existence of a chaotic element in the universe, and others, who find that the Principle of Uniformity is an inalienable constituent of their own minds, and hence of the Whole. There is, indeed, so much parallelism between the notions of uniformity and causality in nature, that,

---

[1] See Chapter v, p. 59.   [2] Cf. Bosanquet, B. (1920), p. 10.
[3] See pp. 82–3 of the present chapter.
[4] See, for instance, Hume, D. (1854), vol. I, *A Treatise of Human Nature*, bk. I. *Of the Understanding*, pp. 120–2, 328–9, etc.; on Hume's conception of causation, see Hibben, J. G. (1910), pp. 90 *et seq*.
[5] Cf. Smith, N. Kemp (1933), pp. 223 *et seq*.

if uniformity is relegated to the rank of an hypothesis, it seems reasonable to give the same status to causality. It is indeed possible that not only uniformity and causality, but also all the other basic assumptions of science, which are treated in general as 'given' and indubitable, might with advantage be transferred to the category of hypotheses; their claim would then be to relative, not to absolute truth.[1]

We have already referred to the recognized fact that the reduction of basic assumptions to the least possible number expedites scientific study.[2] A parallel methodological rule is the so-called 'simplicity postulate', which is sometimes treated as a basic assumption, but which should rather be reckoned as a practical criterion of selection from among 'final inexplicables'. This principle of 'frugality' or 'economy' in thinking can be traced back to classical sources,[3] but it was in medieval philosophy that it played its most conspicuous part. It has been known as 'Occam's razor', since it is implicit in the work of William of Occam (Oakham),[4] who flourished in the early fourteenth century; but the first to express it in the familiar form, *Entia non sunt multiplicanda praeter necessitatem*,[5] appears to have been a seventeenth-century commentator on the Franciscan philosopher, Duns Scotus (b. *c.* 1270), who had made use of the principle. Dante, who was a contemporary of Duns Scotus, also accepted this idea, when he said that "everything superfluous is unpleasing to God and to Nature".[6] Sir William Hamilton in the mid-nineteenth century, drew special attention to this "law of parsimony", as he called it, "which prohibits, without a proven necessity, the multiplication of entities, powers, principles or causes".[7]

---

[1] Cf. Chapter VI, p. 72.   [2] Cf. pp. 80–1 of the present chapter.
[3] Cf. Thompson, Sir D'Arcy W. (1942), pp. 356–7, on the principle of least action.
[4] Carré, M. H. (1946), p. 107; on the use of this principle by Peter Aureolus, who flourished somewhat earlier than William of Occam, see Curtis, S. J. (1950), p. 231.
[5] Thorburn, W. M. (1918), p. 350; see also Harris, C. R. S. (1927), vol. I, p. 167.
[6] Dante Alighieri (1916), *De Monarchia*, cap. I, 14, p. 348: "Et quod potest fieri per unum, melius est per unum fieri quam per plura...omne superfluum Deo et Naturae displiceat."
[7] Hamilton, Sir W. (1852), p. 590.

The history of modern thought seems to show that the simplicity postulate is more convincing to those concerned with the physico-chemical disciplines, than it is to biologists; this is not surprising, since these disciplines are closely connected with mathematics, which deals with a universe simplified by abstraction and rejection; with such a universe the simplicity postulate harmonizes admirably. Sir Isaac Newton accepted the idea that Nature "affects not the pomp of superfluous causes", and he gave, as one of the rules of reasoning: "We are to admit no more causes of natural things than such as are both true and sufficient to explain their appearances".[1] In our own time Sir Harold Jeffreys has shown that "general propositions with high probabilities must have the property of mathematical or logical simplicity".[2] Biologists would not think of disputing the mathematical validity of such a conclusion, but there is no reason to suppose that it is applicable outside mathematics. We have already noticed that the measurable probability of the mathematician falls into a different category from the biologist's probability.[3] Two different terms are really needed for these two conceptions. It may perhaps be suggested that mathematical probability, as an abstraction from probability in the broad sense, is comparable with measurable time, as an abstraction from duration.

It is obvious that in biology, as well as in other fields of non-mathematical thought, the simplicity postulate, like the causal hypothesis, is a convenient instrument of method, in that it reduces the multifariousness of reality to a form with which the human mind can grapple; but it needs cautious use, since it contains an element of that kind of tidiness which simplifies existence by consigning all potentially troublesome documents to the waste-paper basket. We no longer possess that superb nineteenth-century assurance, which led Herschel to state that, the moment we contemplate nature as it is, "we never fail to recognize that sublime simplicity on which the mind rests satisfied that it has attained the truth".[4] On the contrary we have

---

[1] Newton, Sir I. (1803), vol. II, bk. III, p. 160. [2] Jeffreys, Sir H. (1937), p. 7.
[3] Cf. Chapter III, p. 26, footnote 2. [4] Herschel, J. F. W. (1831), p. 361.

learned to-day that it is far from safe to proceed on the naïve assumption that simplicity is equivalent to validity. The biologist especially, faced with the unutterable complexity of living things, cannot but feel that the apparent successes of the simplicity postulate are liable to be achieved by throwing some of the main cargo overboard, under the mistaken impression that it is merely ballast.[1]

A situation that in practice sometimes arises, is that the biologist has before him two alternative hypotheses,[2] one of which is simpler than the other. It is a convenient expedient to adopt the simpler hypothesis provisionally, but, even if it 'works', it should not be accepted at once on the strength of the simplicity postulate, but the more complex hypothesis should also be studied and tested, as there is the possibility that it may rise to a higher level of truth than its simpler rival. We must accept the fact that, if we are led to give preference to an hypothesis on account of its simplicity, this is "non parce qu'elle est la plus vraie, mais parce qu'elle est plus commode et la plus intelligible".[3]

When we come to ask ourselves how it is that we have a conviction of the truth of certain basic assumptions used in science, which are admittedly neither proven nor provable, the answer can be found nowhere but within our own selves. The mind undoubtedly experiences an intense craving for uniformity or intelligibility, and also for simplicity. By sheer natural necessity the intellect, like the body, seeks in the universe for that which is so far conformable to itself that it can be integrated into its very texture. If we may accept the Pythagorean saying that "Man is the measure of all things", it is open to us to believe that such a craving in man, the microcosm, is an indication that in the universe, the macrocosm, there is a corresponding element which the mind searches after and finds. Unfortunately the profound influence of Descartes in biology has tended to neutralize this natural and inherent urge; he impressed a surgical cleavage so persuasively and so effectually upon scientific thought, that the

[1] Cf. Woodger, J. H. (1929), p. 18.　　[2] Cf. Stace, W. T. (1932), pp. 113–14.
[3] Couturat, L. (1901), p. 269, footnote 2; cf. also Thorburn, W. M. (1918), p. 352.

breach he made has never been satisfactorily closed.[1] This unnatural fission sets mind and external nature in opposition to one another, as if they were incommensurables; but it is, in fact, the harmony between the mind and the 'not-self' which renders the 'not-self' comprehensible to man, and those who have seen most deeply into things, have generally found themselves impelled to believe in the existence of this concord. Even the sceptical Hume, in his final revision of the *Dialogues*, added in 1776, a cautious passage in which he accedes to the "somewhat ambiguous, at least undefined proposition, *that the cause or causes of order in the universe probably bear some remote analogy to human intelligence*".[2] In the same year, Sir Joshua Reynolds, approaching life not as a philosopher or scientist but as an artist and man of letters, said that his notion of nature comprehended not only externals, but also the "internal fabrick and organization...of the human mind and imagination";[3] Plotinus, long before, had enshrined a kindred thought in his aphorism—"none walks upon an alien earth".[4]

[1] For a criticism of Descartes's dualism see the Introduction to Ritchie, A. D. (1936); see also Chapter VIII, pp. 98–9, of the present book.
[2] Smith, N. Kemp (1947), pp. 21, 227.
[3] Reynolds, Sir J. (1797), Discourse VII, 10 December 1776, vol. I, p. 136.
[4] Plotinus, *Enn.* v. viii. 4, in Bréhier, E. (1924–38), vol. v, 1931, pp. 139–40; Dodds, E. R. (1933), p. 225, for translation.

BIOLOGICAL ANTITHESES

IN studies of the theoretical aspects of biology, we frequently meet with pairs of ideas set in opposition to one another, because they have some mental connexion, but yet appear to be antagonistic. They are generally called antitheses—a term whose connotation is difficult to define, since it covers concepts which stand to one another in relations which are far from being identical. Sometimes such antitheses are sheer alternatives, of which one or other must be erroneous. The primary antithesis between the conceptions of the universe as a completely ordered Cosmos, and those which see in it at least an element of Chaos, probably come into this category. We need not consider this antithesis here, since we concluded in the preceding chapter that the rational order (or Uniformity) of Nature is that hypothesis which is tested, consciously or unconsciously, in all scientific research. Turning now to those antitheses which are specially the concern of biology, we find that they are not usually simple alternatives, but are of the type in which thesis and antithesis do not necessarily exclude one another. We can detect this feature in such examples as mechanistic explanations of phenomena, contrasted with interpretations in terms of purpose; preformation contrasted with epigenesis; body contrasted with mind; and so on. It may be well to review a few examples of these paired concepts, in order to see if any light can thus be thrown upon their general position in biological thinking. To begin with, we must put aside those apparent antitheses which are based upon misconceptions; an instance is the opposition of *matter* and *form*. When the 'form' of a living thing is discussed as if it were something imposed upon passive 'matter', a mistaken analogy is implied. It is true that the shape of a coin is impressed upon a piece of metal from the outside, so that the ultimate form is not an outcome of the nature of the metal itself, but is something stamped upon it by the agency of man. The living creature, on the other hand, does not consist of passive material upon which form is bestowed by an external cause. The analogy of the coin does not hold, because the metal would have to compose itself spontaneously into coin shape, if it were to be comparable with the fertilized egg cell, which develops, gradually and by inner impulsion, into the form of the mature organism. In plants and animals there is, in reality, no disjunction between the formal and material cause; these causes are divorced only in man's artefacts, and, even here, what we might call a remarriage supervenes in the most consummate works of art, in which the form seems almost to be the self-expression of the medium.

Form and *Function* again, ought never to have been set asunder.[1] The older morphologists do not seem to have separated these concepts with as much rigour as later writers, if we may judge from a remark of de Candolle's, who wrote of the *organized being* as composed of the intimate combination of *functions*, thus showing that to him 'function' and 'organ' must have been interchangeable ideas.[2] E. S. Russell has drawn attention to the integral relation of instinctive behaviour and bodily development, and has shown that they are merely two aspects of phases in the life cycle.[3] In the field of philosophy it is significant that an acute thinker[4] should choose to call his work, *Logic or the Morphology of Knowledge*, explaining that he was borrowing the term 'morphology', not in the sense of a science of external shape which is antithetic to the science of vital function, but rather as referring to that science of life which unites the essence both of morphology (in the narrow sense) and of physiology. Thus, approaching biology in the light of another mode of thought, he recognized clearly that form and function are mere abstractions from the organized whole. A resolution of their apparent antagonism is hinted at when biochemists show us how it may be

---

[1] Cf. Russell, E. S. (1936), p. 9, and Arber, A. (1950), p. 3; on the historic relations of the two concepts, see Russell, E. S. (1916).
[2] Candolle, A. P. de (1813), p. 93.
[3] Russell, E. S. (1934), p. 121.
[4] Bosanquet, B. (1911) (1st ed. 1888), vol. I, pp. 1–2.

possible to understand "the significance of form in terms of chemical composition or metabolic function and vice versa".[1]

Another antithesis which is, from the biological point of view more apparent than real, is that between *subject* and *object*. At the present day we mean by 'subject', him who knows, and by 'object', that which is known—the nature of the 'object' being regarded as unaffected by the thinker whose mind is employed upon it.[2] In all knowledge worthy of the name, this antithesis is resolved ultimately, for we cannot know anything in a full sense until it becomes, as it were, part of the mind. As Spinoza wrote long ago, "it is never we who affirm or deny something of a thing, but it is the thing itself that affirms or denies, in us, something of itself".[3] It is only by transcending the subject-object antithesis that man enters into his heritage, and comes at last to be at home in the universe.[4] Such ideas may seem at first glance remote from the day-to-day work of the biologist, but they have in fact a special relevance to the visual thought employed in his branch of science.[5]

Yet another antithesis, that of *deduction* and *induction*,[6] though a subject of concern to the logician, as a rule troubles the biologist but little. In actual practice, as a student of living things, he often works with the two processes simultaneously, and succeeds in interweaving them happily.[7]

When we turn from the somewhat questionable antitheses, which we have been enumerating, to those paired ideas whose antagonism seems more genuine, we often find that there is a strongly subjective element in their opposition. A cogent

---

[1] Bradfield, J. R. G. (1950), pp. 80–1.
[2] For an account of the change of meaning these terms have undergone between the medieval and modern thought-periods, see Prantl, C. (1855–70), vol. III, 1867, xix, *Duns Scotus*, p. 208; and White, W. Hale, and Stirling, A. H. (1930), pp. vii, viii.
[3] Wolf, A. (1910), Spinoza, *Short Treatise*, pt. II, chap. XVI, p. 109.
[4] Cf. J. B. Baillie, in Hegel, G. W. F. (1931), p. 39.
[5] This point is considered in connexion with biological illustration in Chapter x, pp. 120–21.   [6] See Chapter III, pp. 25, 26.
[7] The unconvincing character of this antithesis has also been recognized from the philosophic side; cf. Adler, M. J. (1927), p. 244, "deductive and analytical processes are involved in any instance of complicated empirical discovery or research"; and Mure, G. R. G. (1932), p. 28: "The fact is that a severance of deduction and induction save in the way of emphasis, is a mark of unsound logical theory."

example is that of the antithesis of *mechanistic* and *teleological* views of the universe; for there is a psychological difference between those biologists who, confronted with some given set of phenomena, are inclined simply to ask, 'How?', and those who are not satisfied with 'How?', but want also to know 'Why?'. The question, 'How?', can often be supplied with a clear-cut answer on materialistic lines, but 'Why?', can find a response only in the much less accessible world of abstract thought. As an instance from botany, the meaning of the distribution of the young leaves at the shoot-apex may be cited. Spatial and bio-chemical explanations can reveal *how* it comes about that the leaf rudiments are placed as they actually are, in relation to one another and to the shoot-apex, but the more elusive question— "*Why* do these spatial and biochemical factors come into play?" —remains untouched. Again, a plant with an abbreviated underground axis tends to develop long leaf-stalks, which lift the leaf-blades well above the level of the soil. This may be interpreted *mechanically* as a growth correlation, the shortening of the internodes encouraging the elongation of the petiole; or *teleologically*, as an adaptation to ensure access of the leaf-blade to the necessary sun and air. In both these botanical examples we are left still asking whether the notion of a purposive reason is, in reality, to be entertained, and in what way such a reason could be related to the 'How?' of mechanism. Again—to take a human instance—when a man writes, the action of his hand, the flow of ink from his pen, and so on, can (in principle, at least) be analysed on purely mechanical lines. If this were done exhaustively, we should know the 'How?' of the writing process, but we should still be in the dark about the reasons which induced the action, and the ends at which that action aimed. Yet the botanical and the human concepts, which we have taken as examples, each had a conspicuous unity before we tried to dissect it analytically; in the initial ideas of leaf distribution and form, and of the process of writing, the mechanistic and teleological aspects existed in harmony. In analysing them out of the primary concept, we have destroyed their relation, and it is this relation which was the guarantee of the original unity.[1] No one can doubt that analysis is an essential tool of biology, but it is also, alas, a lethal weapon. It is arguable that proneness to analysis may be a symptom of panic fear. To man's limited mind, the Whole presents itself as utterly overwhelming; self-defence urges us to dissect it into little bits, with which we can cope individually, and thus to circumvent its terrors. In analysing or abstracting, however, we are tearing asunder something which is homogeneous in its own nature; when we try to reconstruct the whole by putting together the elements thus surgically separated, we find that they have become refractory to being again united. When two different abstractions have been derived from one notion—for instance, the abstractions of mechanism and that of purposive development, from the single concept of the living being—it is impossible to rebuild the integrated concept of the organism as a whole, by juxtaposing these two products of analysis.

So far we have been speaking of antithetic pairs as if they were obvious and easily recognized abstractions from a single whole; but this is by no means always the case. It has been stated by an historian of science that "there are an increasing number of antitheses in the world of our experience which science exhibits no sign of resolving".[2] On the face of it, this dictum may commend itself, and it has, indeed, been maintained that the universe is basically dualistic,[3] so that the opposition of antitheses, or rather of pairs consisting of thesis and antithesis, is fundamental and irreconcilable. Deeper insight is revealed in a view which Coleridge expressed; recognizing the existence of antitheses, he yet was able to see them as the 'coincidence of contraries'.[4] He distinguished, on the one hand, "The *Identity* of Thesis and Antithesis" as "the Substance of all *Being*", and, on the other hand, "their *Opposition*", which is "the condition of all *Existence*, or Being manifested".[5] In less technical terms, we might say that the antithetic pairs, which necessarily

[1] Cf. Bradley, F. H. (1914), p. 193, Appendix to chap. vi: "Is there in the end, such a thing as a relationship which is merely *between* terms? Or, on the other hand, does not a relation imply an underlying unity and an inclusive whole?"
[2] Singer, C. (1931; 2nd ed. 1950), p. viii.   [3] Cf. Sheldon, W. H. (1918).
[4] Cf. p. 110.
[5] Coleridge, S. T. (1818), vol. i, Note to Essay xiii, pp. 155–6.

show as dual in the world of phenomena, to which we have access through the senses, are each fused into a unity in the world of thought, to which the mind has the key. We must be clear, however, about the precise significance of 'fusion' in this sense. It does not mean that dualism is swallowed up in monism, or vice versa, but that thesis and antithesis are included in a further concept, which does justice to both; this concept may be described equally well as a unity which includes dualism, or as a dualism in which unity is implicit. As an illustration we may think, metaphorically, of partners facing one another in a dance; they are *opposite*, but not *opposed* after the fashion of the paired antagonists in a boxing match. The dance partners show as a duality when they are separated, and as a unity when the evolution of the dance brings them together. They offer an example of what may be called a bipolar unity, representing simultaneously both difference-in-identity, and identity-in-difference. We are reminded of Croce's interpretation of Hegel's dialectic as "a thinking of reality as at once united and divided";[1] and of the dynamic view of "that rhythm of tension and release which Goethe...felt to be the very pulse of the universe", and which he visualized as polarity.[2] The biologist must, indeed, often feel a consciousness of such essential rhythm in his own work. George Eliot entered into the researcher's standpoint when she made Lydgate voice the recognition that "there must be a systole and diastole in all inquiry", and that "a man's mind must be continually expanding and shrinking between the whole human horizon and the horizon of an object-glass".[3]

It is conceivable that, in the mental world, the idea of bipolarity may be out-distanced, since unified thought might be visualized as *multipolar*; but the mind clings to the simpler bipolar concept, ingrained into our thinking, perhaps in correlation with the dual symmetry inherent in man's bodily structure. In the relation between bipolar and (possibly) multipolar thought, a parallel may be traced to Spinoza's doctrine of the Infinite Attributes of the Whole, of which, however, two only—

---

[1] Croce, B. (1915), pp. 19–20.   [2] Wilkinson, E. M. (1949), p. 309.
[3] Eliot, G. (1871–2), vol. IV, bk. VII, chap. LXIII, p. 6.

Extension and Thought—are within the compass of the human understanding, which thus reduces the Infinite to bipolarity. However this may be, for the general purposes of the biologist, bipolarity is an adequate concept, symbolizing something which meets him recognizably at every turn. The fact that each organism is both a unity intrinsic to itself, and also an integral part of the nexus which is the Whole,[1] informs it with a basic duality. The organism is thus polarized in one directional sense towards the Whole, and in the other directional sense towards the core of its own innermost being. A second somewhat different form of polarity is that between a man's mind and his body. This may be regarded as a special case of Descartes's antithesis of thought and extension.[2] Descartes was of opinion that 'body' and 'mind' were contrasted and entirely separate entities,[3] temporarily conjoined during the earthly life; he seems, indeed, to have been the first philosopher to maintain that there is a complete and essential heterogeneity between mind and body.[4] In order to justify this rigid separation, he had to limit the term 'mind' to the reasoning faculty, which he held to be denied to other living things.[5] This hard-and-fast limitation of the intellectual field has an inhibiting effect on interpretations. It is more enlightening to define the mind in a broad sense, as annexing other parts of our thinking, as well as the restricted region which employs merely abstract reasoning. The term, 'mind', should include, for instance, the artist's and biologist's visual thought, as well as other forms of mental activity associated, not only with the roof-brain, but also with the senses. Unlike Descartes, Spinoza treated the concepts of

[1] This statement can be provisional only, since it depends upon the hypothesis of the Uniformity of Nature.

[2] Descartes, R. (1644), p. 20, pars I, princ. LIII (marginal title), "Cujusque substantia unùm esse praecipuum attributum, ut mentis cogitatio, corporis extensio". Translation, Haldane, E. S., and Ross, G. R. T. (1911–12), vol. I, p. 240.

[3] Descartes, R. (1641), Synopsis, p. 3, "substantiae diversae, sicuti concipiuntur mens et corpus, esse revera substantiae realiter à se mutuò distinctas". Translation, Haldane, E. S., and Ross, G. R. T. (1911–12), vol. I, p. 141.

[4] Cf. Coleridge, S. T. (1817), vol. I, p. 128; and Coburn, K. (1951), pp. 63–4 (Coleridge's unpublished sketch for an Essay on the Passions).

[5] [Descartes, R.] (1637), p. 58; (1947), p. 58. Translation, Haldane, E. S., and Ross, G. R. T. (1911–12), vol. I, p. 117.

body and mind as both referring to the same reality; this reality was, however, held to assume an entirely different character according to the 'attribute'—extension or thought—under which it was considered. (As a crude comparison, illustrating the nature of 'attributes', we may visualize the different aspects which the same landscape presents, according to whether it is seen at noon or midnight.) Whereas Descartes's view of body and mind offers an unresolvable antithesis, on Spinoza's theory, the antagonism is in part overcome.[1] Perhaps we may modify Spinoza's view so far as to think of the body as the individual considered under the aspect of multiplicity, since the body may be seen, from one standpoint, as consisting of component parts whose structural and functional variety beggars description; we should then regard the mind as the individual contemplated under the contrasting aspect of unity;[2] but the realization of relatedness may offer a bridge between these conceptions. Though we may choose deliberately to think of the elements of the body in isolation, this can be done only by a process of artificial abstraction; actually they are intimately and fundamentally interwoven, even when this is not, at first glance, obvious. Bodily multiplicity, in its ultimate organic relatedness, *becomes* unity, and this unity *is* the mind-body individual. Far from being a bare unity without variety, it comprehends in itself all multifariousness and all the strands of interrelatedness which connect the wealth of discrete details. Multiplicity and change exist on the background of space-time, while unity and changelessness are free from the limitations of duration and place, since, in their very nature they belong to the eternal things. The body is the individual considered *sub specie temporis et loci*, while the mind is the same individual seen under the opposed aspect of eternity; but in order to relate these two aspects, we need to go beyond this limited and one-sided conception of eternity, by enriching it to include all multifariousness and all temporality.

---

[1] On Spinoza as conciliator, see Wolf, A. (1922).

[2] Descartes, R. (1641), Med. vi, p. 109, "nam sanè cum hanc considero, sive meipsum quatenus sum tantùm res cogitans, nullas in me partes possum distinguere, sed rem planè unam et integram me esse intelligo". Translation, Haldane, E. S., and Ross, G. R. T. (1911–12), vol. i, p. 196; "When I consider the mind...I...apprehend myself to be clearly one and entire."

7-2

A slightly different expression of the same point of view is found in Leibniz's suggestion that "Minds act in accordance with the laws of final causes.... Bodies act in accordance with the laws of efficient causes".[1] To regard the living thing under the aspect of efficient causes, is equivalent to regarding it in the context of temporal and spatial succession—that is to say, as part of 'extension', if we may broaden this term to include time as well as space. So looked at, the living thing is 'body'. On the other hand, if we think of it in terms of formal and final causes alone, we are seeing it from the standpoint of eternity—using this word, however, in the narrow sense in which it is antithetic to space-time. The individual is then regarded simply as 'mind'; but we can reach beyond these contrasting positions, if we realize that causality, in its ordinary meaning, can only be thought in connexion with time. Whether we regard this causality as material-efficient (mechanistic: 'How?') or as formal-final (teleological: 'Why?'), depends on the time-sequence, and the 'sense' in which we pursue its direction. If, on the one hand, we pass mentally from past to future, the material-efficient causes disclose themselves; if, on the other hand, we start with the future and think back to the past, the formal-final causes are revealed. If we eliminate the reference to time, we can synthesize these two types of cause, seeing them as complementary aspects of one whole, and we thus reach the concept of the mind-body individual, seen *sub specie aeternitatis*—eternity here being given its fullest connotation.

These considerations about body and mind show that their opposition may be treated as a special case of a wider biological antithesis—that between the mechanistic and vitalistic theories of the animate world. The *mechanist*, starting from the physico-chemical standpoint, interprets the living thing by analogy with a machine. The *vitalist*, on the other hand, supposes a guiding entelechy, which summons order out of chaos; he thus

---

[1] Leibniz, G. W. (1930), sect. 79, pp. 184–5: "Les âmes agissent selon les lois des causes finales par appétitions, fins et moyens. Les corps agissent selon les lois des causes efficientes ou des mouvements. Et les deux règnes, celui des causes efficientes et celui des causes finales, sont harmoniques entre eux." Also, on Aristotle's four causes, see Arber, A. (1950), p. 199.

adopts a dualistic attitude. The elements of truth in both these views are recognized, and their opposition is resolved in the *organismal* approach to the living creature. This approach is conditioned by the belief that the vital co-ordination of structures and processes is not due to an alien entelechy, but is an integral part of the living system itself.[1] This notion has kinship with Spinoza's doctrine of the *conatus*, or urge of the creature towards self-maintenance—an urge which he equates with the actual essence of the thing itself,[2] and with its very life.[3] Organismal theories differ from the vitalistic in much the same way that Aristotle's view of the 'forms', as being inherent and immanent in things, differs from Plato's conception of these 'forms' as having a separate existence in a supersensible world.

Vitalism and mechanism, as one of the antitheses of general biological thought, may be correlated with epigenesis and pre-formation,[4] as an antithesis in the study of embryology. William Harvey in the seventeenth century followed Aristotle in up-holding *epigenesis*; he believed, that is to say, that the organs were gradually and successively differentiated from the undifferentiated embryo; these organs were not originally present in miniature, but the germ possessed the power to form them anew. In the same century there was a revival of the contrasting Hippocratean[5] theory—that of *preformation*, according to which the apparently undifferentiated embryo had within itself "a complicated machine-structure corresponding with the visible structure of the adult",[6] which needed only to become explicit. Biologists as a rule reck little of formal logic, but it may be noticed that there is a close, albeit unconscious, parallel between

[1] Russell, E. S. (1930), pp. 190, etc.; Bertalanffy, L. von (1933), pp. 177, etc.; Wheeler, R. H. (1935), pp. 344, etc.

[2] It has been suggested that the essence of individual life is described more accurately as, in Aristotelian terminology, the "drive towards the actualisation of potentialities, to which self-maintenance is a means", Russell, E. S. (1945), p. 191.

[3] [Spinoza, B. de] 'B.D.S.' (1677), *Ethices*, pars III, prop. vii, p. 102; Vloten, J. van, and Land, J. P. N. (1882–3), vol. II, p. 487 (Spinoza's *Cogitata metaphysica*, pars II, cap. vi); cf. also Arber, A. (1950), p. 77.

[4] Cf. Nordenskiöld, N. E. (1950), pp. 117–18, 170, etc.; Russell, E. S. (1930), pp. 26–7.

[5] Cf. Mure, G. R. G. (1932), p. 100; it thus seems that the preformation theory had an earlier origin than Russell suggests.

[6] Russell, E. S. (1930), p. 27.

certain contrasting views of logical inference and these two contending views about embryology. The idea that the adult is preformed in the embryo, and needs merely, as it were, to unfold, may be compared with the view that formal inference does not lead to new discoveries, but only exposes knowledge which was already present in the premises. On the other hand, epigenesis corresponds to that view of inference which regards it, not as a mere analysis of premises, but as creative, in the sense that the relations which it brings light to were not, strictly speaking, present originally in the premises, but are emergent from them.[1]

As vitalism and mechanism are synthesized in the organismal theory, so epigenesis and preformation are, to some extent, synthesized in certain modern theories of embryology, which, while mainly stressing epigenesis, yet do not lose sight of the physico-chemical mechanism of development.[2]

In plants, unlike animals, within the developmental history of the individual, we may meet with two antithetic stages—a vegetative phase, which is essentially juvenile, and a reproductive phase, which is characteristic of maturity. Their antagonism is revealed, for instance, in the excessive vegetative growth of many aquatics, which tends to replace sexual reproduction;[3] on the other hand, in land plants, flowering may be associated, as in certain bamboos,[4] with the death and destruction of the vegetative parts. In the asexual phase, the urge to individual self-maintenance is dominant, while, in the reproductive phase, this urge is overcome by the urge to race-continuance. We have here what seems at first sight to be an insoluble antagonism, but we discover a hint as to how this opposition may be transcended when we remember that the individual is bipolar—in one aspect it is unique, but in the opposite aspect it represents the race. Reproduction *is* self-continuance, interpreted in the broad sense that redeems the maintenance of self from egoism.[5]

[1] On these two views of inference, see Walsh, W. H. (1947), p. 46.
[2] These are discussed in Russell, E. S. (1930), chap. VI, pp. 76–94; on Y. Delage, as combining these interpretations, see p. 79.
[3] Cf. Arber, A. (1920), pp. 210–26.
[4] Cf. Arber, A. (1934), pp. 100–1.
[5] On reproduction as the goal of self-continuance, see Arber, A. (1950), pp. 20, 78.

A further antithesis, of a different type, is that between the organism and its environment. It has been suggested that this antithesis is a genuine and irremovable one,[1] and emphasis has been added to this antagonism by stressing the power of the living creature to maintain its organization in the teeth, as it were, of the environment.[2] On the other hand, a contrary view has been recognized—namely, that the reaction of the higher animals to their environment as a whole is an involvement, as we might call it, rather than an opposition; there is, indeed, a capacity in the central nervous system which enables the animal "to react to a unified 'world' instead of to a series of discrete stimuli".[3] The same line of thought has been extended by a physiologist,[4] who has urged that co-ordination is as fully maintained between organism and environment as between the parts of the organism itself, and that we cannot, even mentally, separate the phenomena of life from those of the environment.

Hitherto we have been considering the antitheses which are most obviously bound up with biology, but, as a sequel, it may be worth while to glance at the more general question of the antithesis of the One and the Many, which is indeed fundamental for metaphysics. This far-reaching antithesis has, in fact, a special relevance to the work of the scientist, since it is related to the hypothesis of the Uniformity, or Ultimate Oneness, of Nature, in which the Many find union. The contrast of the One and the Many was stressed by the pre-Socratics; in the *Parmenides* Plato reviewed earlier opinions, and analysed the various senses in which 'the One' can be understood.[5] When we consider the One as a Whole of Parts, it is *One* if we emphasize the word *Whole*, while it is *Many* if we focus attention upon the word *Parts*.[6] Various thinkers have believed that, broadly speaking, it is to the intellect that we owe the recognition of identity or unity, while the senses give us diversity;[7] it may, indeed, be held that the unity of consciousness *is* the mind

---

[1] Woodger, J. H. (1929), p. 332. Woodger's discussion of this and other antitheses should be compared with their treatment in the present chapter.
[2] Robson, G. C., and Richards, O. W. (1936), pp. 352–3; Young, J. Z. (1938*b*), p. 515. [3] Young, J. Z. (1938*a*), p. 192. [4] Haldane, J. S. (1935), pp. 45–6.
[5] Cornford, F. M. (1939). [6] Taylor, A. E. (1918), p. 611–12.
[7] Cf. Meyerson, É. (1931), vol. II, pp. 574, 579 *et passim*.

of man.[1] Those who lean to synthesis, and in whom non-sensuous contemplation predominates, thus look instinctively for the identity and unity of the Whole, while those of an analytic turn, for whom the senses and perceptions offer the primary road to reality, are more alive to the innumerable many-faceted phenomena of the world around them, and are less inclined to look for an abstract unity. This distinction to some extent corresponds to that between the votaries of philosophy and of natural science, but it is far from absolute. In its application to students of different disciplines, it is merely an example of those contrasting inborn trends towards one or other pole of the world of experience, which differentiate personalities. This temperamental contrast is exemplified in the history of biology in the grievous and protracted polemic, between Cuvier and Geoffroy Saint-Hilaire, about the significance of structural relations in the animal kingdom.[2] Cuvier's clear intellect, reinforced by immense factual knowledge based on countless dissections, led him to the appreciation of differences rather than resemblances, so that he regarded as unbridgeable the gaps between the main types of animal organization. On the other hand, Geoffroy's mentality—in its mysticism, Teutonic rather than Gallic—was drawn irresistibly towards comprehensive views, centering in the idea of universal unity; he thought of 'l'Animalité' as "être abstrait, qui est tangible par nos sens sous des figures diverses".[3] Actually there was some truth in the views of both men, and the philosophic basis of their disagreement was delusive; for to create a hard-and-fast cleavage between the senses, as the medium for the multifariousness which appealed to Cuvier, and the mind, as the source of the unity for which Geoffroy craved, is to mistake a difference, which has only relative validity, for an absolute distinction.

Of the multiplicity of phenomena, we have at least hints and samples of a mental picture, but the idea of the unified Whole[4]

[1] Cf. Caird, E. (1889), vol. I, p. 19.
[2] For an account of this controversy, see Nordenskiöld, N. E. (1950), pp. 341–3; and for a detailed and documented study of Cuvier, with incidental reference to Étienne Geoffroy Saint-Hilaire, see Daudin, H. (1926), vol. II, pp. 71–109.
[3] Geoffroy Saint-Hilaire, É. (1830), p. 22.
[4] For a stimulating historical study of science and monism, see Wightman, W. P. D. (1934).

is more recondite, and defies pictorial thinking. The revelations of anatomy and physiology offer, however, some faint suggestion as to how Wholeness may be built anew by the brain out of the heterogeneous material which it receives. In the case of the higher animals, the concentrated central nervous system renders possible the correlation of perceptions from the different senses, as well as other co-ordinations, thus providing a mechanism which is capable of organizing functional multifariousness into unity.[1] It is, moreover, a possible view that this organization is implemented by the warfare of opposites. The unity of the tissues of the body may be, not a unity of simple peace, but an armed neutrality, in which equipoise is secured by stresses and strains which balance one another.[2] This conception, of a unity based on conflict, is applicable to organs as well as to tissues. The final harmonious symmetry of the plant body, for instance, may take its origin in competition between the members, each shoot-generation showing an urge to dominate, or even to replace its parent; the rhythmic patterns of various kinds of sympodial growth arise in this way.[3]

When we try to trace the concept of wholeness, as it develops in a man's mind, we realize that, as an infant, his vision of the surrounding world possesses a certain primitive unity, since, in a sense, it forms a whole, not yet discriminated into components. At the opposite pole is the kind of unity achieved by mature thought, in which fully analytical observation of individual things, and the differentiation of individual ideas, has been followed by a synthesis which has reconstructed unity from diversity. Between these two poles—the first, unconscious, and the second, self-conscious—lies the whole developmental sequence of the intellectual life.[4]

If the antithesis of the One and the Many is to be resolved effectively, due value must be given to both terms. Unfortunately, since the thinkers most interested in this synthesis

---

[1] On this subject, see Young, J. Z. (1938a).   [2] Cf. Roberts, M. (1920).
[3] Cf. Arber, A. (1950), chap. vii, pp. 93 et seq.
[4] For an account of unity which suggests this view, see Schiller, F. C. S. (1931), pp. 281–2; cf. also Coleridge, S. T. (1895), p. 53 (11 December 1803): "The dim intellect sees an absolute oneness, the perfectly clear intellect knowingly perceives it. Distinction and plurality lie in the betwixt."

generally possess that reflective habit of mind which leans to the One rather than to the Many, they tend to turn an unseeing eye upon the multifariousness of existing things. Goethe, for instance, was disposed to grasp too precipitately at Oneness, and to shrink from the conception of detailed Manifoldness.[1] It is true that, to see things in their reality, we have to apprehend them *sub specie unitatis*,[2] but this unity must be that Unity-in-plurality, or Singleness-in-complexity,[3] which includes multiplicity while transcending it, and which may be distinguished as Totality.[4] Indian thought seems to have recognized the need to pass beyond the naïve conception of Oneness, more fully than Western philosophy,[5] though Heraclitus, writing before the birth of Socrates, realized that the Unity of all things is not simple Oneness, but is the tension of opposites; unity in the manifold; the harmony of strife; order within change.[6] Other European thinkers have followed him. For instance, in the Renaissance period, when Bruno said: "It is Unity that doth enchant me. By her power I am...quick even in death",[7] he was visualizing the Infinite One, not as bare Oneness, but as containing and enfolding the manifold conclusions of *scienza naturale*.[8]

When we come more closely to grips with the relation of the One and the Many, we are confronted with the question as to whether the Many are to be conceived as emanations from the One, or whether the idea of the Many has priority, as the source from which that of the One is derived. In other words, we have to consider whether there is more truth in the Platonic tradition that the Whole precedes the Parts, or in the Democritean view that the Whole is achieved through the Parts. We can only

[1] Cf. Arber, A. (1946a), pp. 80, etc.
[2] Cf. Muirhead, J. H. (1931; 2nd imp., 1939), p. 417.
[3] Cf. Joachim, H. H. (1948), p. 217.
[4] Cf. Kant, I. (1902–38), Bd. III, 1904, *Kritik der Reinen Vernunft*, 2nd ed. 1787. Elementarlehre, Th. II, Abth. I, Buch I, Hauptstück I, pp. 93 and 96. Categories of Quantity—Einheit (Unity), Vielheit (Plurality), and the category representing the previous two in combination—Allheit or Totalität (Totality).
[5] Cf. Guénon, R. (1945), p. 154; Hiriyanna, M. (1949), pp. 162, etc.
[6] Patrick, G. T. W. (1888), p. 618, or (1889), p. 59; cf. Burnet, J. (1920), p. 143.
[7] Singer, D. W. (1950), p. 229; Bruno, G. (1923–7), vol. I, 1925, p. 270, "una che m'innamora; quella per cui son...vivo ne la morte" (1584).
[8] Singer, D. W. (1950), p. 101; Bruno, G. (1923–7), vol. I, 1925, p. 144; "Quivi, come nel proprio seme, si contiene ed implica la moltitudine de le conclusioni della scienza naturale" (1584).

accept the Platonic position if we suppose that priority in being and priority in time are two different things, so that, in the order of reality, the Whole precedes the Parts.[1] Broadly speaking, this is the view that the metaphysician accepts. He starts from Unity, and deduces the multiplicity of existing things—that is to say, he is committed to the perilous, if not impossible, task of deriving the Many from the One. This is, in fact, what he is trying to do, even if, like Hegel, he seems to reach the Absolute only as the culmination of his dialectic. It is characteristic of the biologist, on the other hand, to start with the multiplicity of the world of living things, and to generalize his experience step by step, until he eventually comes within sight of Unity. Rather than Plato, he is following Aristotle, who seems to have held that perception itself contains an element of the universal, and that thus, from the perception of particular things, we can make the transition to higher and higher grades of universality.[2] When we grasp the universals implicit in the particulars, we may be said to be proceeding from the Many towards the One. The biologist's conception of the One, achieved in this way through the study and apprehension of the Many, is summed up in the motto engraved on Richard Owen's signet ring, which symbolized the unity of plan in the vertebrate skeleton—"The One in the Manifold".[3] Such a conception may perhaps claim a greater richness of intellectual content than the metaphysician's intuitive postulation of the same unity. These two modes of approach have, however, more in common than may appear at first glance. The metaphysician, who takes the Ultimate One as the basis of his world picture, has at least a subconscious awareness that this unity is something which he has, in actual fact, disengaged from the endless plurality of the universe. The scientist, on the other hand, would never have the heart to struggle through the successive painstaking stages in his laborious inductions, if he were not moved thereto by an instinctive underlying bias towards a belief in the Uniformity of Nature, and hence in the unity of all things.

[1] Cf. Muirhead, J. H. (1931; 2nd imp., 1939), p. 418.
[2] Ross, W. D. (Sir D.) (1937), cf. pp. 54–5.   [3] Owen, R. (1894), vol. I, pp. 387–8.

CHAPTER IX

ANTITHESES AND DIALECTIC

THE irreducible opposition of certain biological concepts may sometimes be ascribed to their relevance to different levels of thought—those limited 'universes' of ideas, within the conceptual universe as a whole, which de Morgan long ago distinguished.[1] In the physico-chemical region of biology, for instance, one case may at the same time be susceptible both of a mathematical and also of a biochemical interpretation. For example, Dormer's[2] demonstration that the arrangement of the parts in certain flower-buds is an expression of geometric necessity, does not preclude a biochemical reading of this aestivation, according to which the position of the rudiments of floral parts is held to be regulated by chemical inhibitions exerted by previous members. Such mathematical and biochemical theories may eventually find a synthesis on a plane higher than either. Physiological and morphological problem-solutions, also, may appear very different from one another, but they can sometimes be seen as together constituting a 'two-level' theory. Wardlaw[3] has suggested, tentatively, on the basis of experimental work of a delicate and convincing kind, that the dorsiventral symmetry of leaves may be related, *in part*, to the inhibiting influence of the apical cell-group of the shoot. This affects the faces of the rudiments nearest to itself, checking full radial development, and thus inducing the characteristic flat leaf structure. Other rudiments, however, which are more remote from the centre of inhibition, are not prevented from forming themselves into radial structures, and becoming shoots. A morphogenetic theory of this type in no way conflicts with Casimir de Candolle's purely morphological idea that the leaf is a partial shoot, reduced to dorsiventrality by the atrophy

---

[1] de Morgan, A. (1847), p. 41.   [2] Dormer, K. J. (1948), pp. 653–4.
[3] Wardlaw, C. W. (1950), p. 16.

of its adaxial face;[1] each of these theories is valid on its own plane.

When we pass in review the antitheses which confront the biologist, we see that, in general, they cannot be called examples of strict alternatives of the 'either...or...' type. They belong, to the 'both...and...' category, for they tend, as we have shown, to be pairs of partial truths, which seem opposed because of their incompleteness, but are capable of being brought together in a statement of a higher order, approaching a degree nearer to the full truth than either of them does singly. This principle was one which Kant recognized as basic to his own philosophic procedure. In a letter of 1771 he wrote: "You know that I do not approach reasonable objections with the intention merely of refuting them, but that in thinking them over I always weave them into my judgments, and afford them the opportunity of overturning all my most cherished beliefs. I entertain the hope that by thus viewing my judgments impartially from the standpoint of others some third view that will improve upon my previous insight may be obtainable".[2]

This may seem to be a commonplace sentiment, to which every single-minded research worker would, in theory, subscribe; but there is more in it than meets the eye. It was the principle underlying these remarks of Kant's which formed the skeletal system of the Socratic type of argument;[3] moreover, after Kant, Hegel[4] developed it as the basis of his whole dialectic, which consists in the progressive transcending of

---

[1] On this theory and its modern applications, see Arber, A. (1950), chaps. VI–VIII, pp. 70–131.

[2] Kant, I. (1902–38), vol. x, 1900, pp. 116–17. Letter to Marcus Herz, 7 June 1771: "Dass vernünftige Einwürfe von mir nicht blos von der Seite angesehen werden wie sie zu wiederlegen seyn könten sondern dass ich sie iederzeit beym Nachdenken unter meine Urtheile webe und ihnen das Recht lasse alle vorgefasste Meinungen die ich sonst beliebt hatte über den Haufen zu werfen, das wissen sie. Ich hoffe immer dadurch dass ich meine Urtheile aus dem Standpunkte anderer unpartheyisch ansehe etwas drittes herauszubekommen was besser ist als mein voriges." Translation from Smith, N. Kemp (1923), p. xxii. For an expansion of this Kantian principle, see Caird, E. (1889), vol. I, pp. 7–8.

[3] On the Socratic type of argument from this standpoint, see Mure, G. R. G. (1932), pp. 28–9.

[4] Hegel acknowledged his indebtedness to the triplicity of Kant's system; see Hegel, G. W. F. (1931), p. 107; J. B. Baillie's Introduction to this translation summarizes the general Hegelian attitude.

antitheses.[1] In each triad of his argument, thesis is opposed to anti-thesis, and then these two are fused into an emergent synthesis, which is nearer the truth than either of its components. This is not, however, the end. Hegel shows that in each case the antithesis is *implied by* the thesis; moreover, the synthesis, though itself on a higher plane than either thesis or antithesis, is still an imperfect truth, and contains in its core the seeds of another opposition. This degrades it into the position of a thesis, from which a further antithesis emerges. This new thesis and antithesis are resolved on a higher level; but the synthesis so obtained shows im-perfection in its turn, thus descending to the status of a thesis. The scheme is hence repeated time after time. In this way, by the continuous movement of his dialectic, Hegel passes from Being, his first thesis, until he reaches the final synthesis of the Absolute, in which the sequence terminates.

Very few philosophers to-day would accept Hegel's scheme of things entire, but most thinkers might grant that he did an essential service in securing a renewed recognition of the 'coinci-dence of contraries'—the fact that it is possible for 'A' to be 'both A and not-A'. This idea, which has suffered periods of dormancy and of recurrence throughout the ages, was stressed by Nicolas of Cusa in the fifteenth century, and, later, by Giordano Bruno, under Cusan inspiration. Bruno held, for instance, "that contraries do truly concur; they are from a single origin and are in truth and substance one".[2] It is clear that, if such a point of view be accepted, the pursuit of truth becomes an infinitely more complex matter than if we are contented with the naïve notion that 'A is either A or not-A'. The coincidence of contraries undermines the simple 'either... or...' solution of any knotty problem, because no continuous linear sequence of thought can do justice to truth, if its multi-faceted character is appreciated fully.

[1] Among the many books which aim at enlightening the student about Hegel's dialectic, Stace, W. T. (1924), is particularly useful to the amateur. McTaggart, J. McT. E. (1922), will be found most stimulating, though it must be borne in mind that it sometimes stresses the author's own views rather than those of Hegel.

[2] Singer D. W. (1950), p. 101; Bruno, G. (1923-7), vol. 1, 1925, p. 144, "s'apportano gli segni e le verificazioni per quali gli contrarii veramente concor-reno, sono da un principio e sono in verità e sustanza uno" (1584).

The core of Hegel's dialectic seems to lie in a graded approach to reality, implemented by what might be described as 'tacking' from side to side. This process to some extent corresponds to the unravelment of a problematic topic by means of dialogue. Indeed, the method of discourse or disputation between different interlocutors, or between the mind and itself, permeates the whole process of philosophy. The expression, 'discursive thought', means essentially thought expressed in discourse, while 'dialectic' is that which belongs to discourse or argument. Plato's dialogue technique was completely suited to express the varied nuances of a subject; to mark their differences; and to suggest their possible reconcilement. The structure of the Platonic dialogues represents, in fact, a transition to the full dramatic mode—that mode in which Shakespeare formulated the innumerable aspects of his own vision. In no other way can incompatibilities and harmonies find so effectual a medium. Every speech, written or spoken by one person, suffers under the disadvantage of being a simple sequence of words, linear in time or space; the dramatic form, on the other hand, gives opportunity for a number of sequences instead of one. Not only Plato, but many later thinkers, have found that dialogue supplied them with exactly the framework their thought needed. It serves to reveal the multifarious facets of reality, as seen through the windows of differing personalities; like Shelley's "dome of many-coloured glass", it tempers the "white radiance" which the eye of man cannot face in its purity. The dialogue form served Bruno well in his high speculations, for instance in his work *On the Infinite Universe and Worlds* (1584).[1] Descartes, also, in his study of the search after truth by means of the light of nature,[2] put what he had to say in the form of a discussion carried on between three men of varied character and attainments. British writers have used the same convention with striking effect; for instance, Robert Boyle in *The Sceptical Chymist*

[1] *De l'Infinito Universo e Mondi*, in Bruno, G. (1923–7), vol. I, 1925, pp. 267–418; translation in Singer, D. W. (1950); for Bruno in English see also Greenberg, S. (1950).
[2] *La recherche de la vérité par la lumière naturelle*, in Descartes, R. (1897–1910), vol. X, 1908, pp. 495–527; translation in Haldane, E. S., and Ross, G. R. T. (1911–12), vol. I, pp. 305–27.

(1680),[1] George Berkeley in *Three Dialogues between Hylas and Philonous* (1713),[2] and David Hume in *Dialogues concerning Natural Religion* (1779).[3] It has been suggested[4] that the marked recurrence of the dialogue form in the eighteenth century may be associated with that era's character as the Age of Conversation; in this respect it faintly recalls the Athens of the time of Socrates, in which discourse with their fellows was a natural mode of self-expression for freemen. Among the Greeks, even when dialogue was not being used ostensibly, philosophic writing might show the impress of this form. Some of the incompatible statements in Aristotle's treatises[5] may be compared with the different views of the same subject-matter, which Plato assigns to different characters in the dialogues, thus avoiding the personal inconsistency which Aristotle disregarded. Kant, again is far from being steadily consistent throughout his work; but posterity may be grateful to him for deliberately giving varied but progressive expositions of his problems, almost as though they were seen through the minds of different people.

Written correspondence may be grouped with dialogue, as offering, from its to-and-fro quality, some of the same advantages; this was evident in the pre-modern period when, for men of learning, communication by letters, which were then passed from hand to hand and discussed, held the place of the present exchange of off-prints from journals. The questions and criticisms of Spinoza's correspondents, for instance, sometimes give a fresh orientation to the subjects under discussion.

That Plato expressed his philosophy in dialogue form, enabled him to refrain from any final *ex cathedra* pronouncement, so that it cannot at the end be said with decision that Plato himself thought this or that. On the other hand, in science to-day, an author is expected to adopt, and to outline firmly, a single definite view, expressible in as few words as possible, and to

---

[1] Boyle, The Hon. R. (1680).   [2] Jessop, T. E. (1949), pp. 147–263.
[3] Hume, D. (1854), vol. II, pp. 408–540; Smith, N. Kemp (1947).
[4] Jessop, T. E. (1949), p. 155.
[5] Those of Aristotle's works which were in dialogue form have not survived; see Mure, G. R. G. (1932), pp. 254–5.

conclude his treatise with a neat summary, readily comprehensible, and easy to remember. All this undoubtedly means a gain in clarity, but it generally involves a corresponding sacrifice of truth. For Renan's query[1]—"Qui sait si la finesse d'esprit ne consiste pas à s'abstenir de conclure?"—is as relevant to biology as to any other branch of the intellectual life. The very nature of scientific conclusions is to be inconclusive, and it is impossible for any biologist to hope, for one moment, that his own pronouncements will remain the last word. On a scientific question no deliverance from any one individual can ever be permanent and decisive, for, in this as in other regions of thought, one man's personality could never provide an adequate medium of formulation for a complete and harmonious synthesis. At first sight this may seem an inhibiting idea, but actually it contains a hint of encouragement. When the biological notions of the present day are envisaged broadly, they show as a heterogeneous medley, often groupable in such antithetic pairs as those at which we have been glancing; but this medley—though it may revolt the mind that craves for a rigid system—perhaps offers *in toto* a closer approximation to the truth than can be afforded by any clear-cut scheme attempting universality. When we are thinking about the work of artists, we do not complain because each of the individual painters, even in one period, gives us, from his own particular angle, his own kind of truth; if the points of view of these artists could be somehow amalgamated, the upshot would be the utter sterility of a composite photograph. Science and art are far more closely related than is generally realized, and it is hard to see why biologists should be forced to accept one unvarying communal outlook, to the neglect of all other possible approaches, while artists are free each to take his own individual line.

Though dialogue and its variants are of the utmost value in interpreting and criticizing thought, the same cannot be claimed for polemical controversy, which, as John Norris of Bemerton said in the seventeenth century, is "a thing of great Labour and

---

[1] Renan, E. (1852), p. v; for a study emphasizing the essential inconclusiveness of dialectic, see Adler, M. J. (1927).

but little profit. . . a Litigious wrangle, proceeding. . . from men's mistaking or misstating the thing in Question, from misunderstanding of the Point, of themselves, and of one another".[1]

Controversy, indeed, in the long run, offers as many certain evils, to set off against its doubtful benefits, as the so-called Holy Wars with which history is chequered. Modes of argument expressible in military metaphors need vigilant and critical scrutiny. The experience of most biologists cannot but lead them to feel that the best course is to put their energies into the constructive search for truth, in the confidence that fallacious views, if left to themselves will die of inanition, whereas the stimulus of combat may revive them. The polemical spirit encourages the cowardly and wasteful practice of attempting to refute a system by attacking its feebler aspects, and ignoring whatever truth other facets might reveal; as Mill[2] said, long ago, it is the most reasonable rather than the absurdest forms of a wrong opinion with which one ought to grapple. If this is done, some truth may be wrung even from theories that at first sight seem unpromising; but this can be achieved only if these theories are grasped, as Jacob grasped the Angel, with the determination, "I will not let thee go, except thou bless me". Aristotle understood this. He has been described as seeking to recommend any solution of a problem, which he has to offer, by showing the extent to which, in its light, the opinions opposed to it could be interpreted as partially true, rather than as wholly mistaken.[3]

[1] Norris, J., *Spiritual Counsel; or, the Father's Advice to his Children*, sect. 43, quoted in Powicke, F. J. (1894), p. 24.
[2] Mill, J. S. (1843), vol. II, p. 539; see also Renan, E. (1852), p. 105.
[3] Stocks, J. L. (1938), pp. 93-4.

CHAPTER X

THE MIND AND THE EYE

MANY of the antitheses of biology can be traced to modifications of the primary antithesis between *Thought* and *Extension*—to use the terminology of Descartes and Spinoza. If we concentrate attention, for example, upon the *Unity* of thought, as compared with the *Multifariousness* of the things which compose the extended world, we see that, in its widest development, this antithesis expands into the general relation of the *One* and the *Many*, which again can be narrowed to an individual instance in the antithesis of *Intellect* and *Senses*. In the work of the biologist, visual impressions play a more significant role than those from the other sense organs; we may thus treat the antithesis of *Mind* and *Eye* as epitomizing the much broader subject of the relation of the intellect to the senses in general. It will be convenient to consider this antithesis, first, from the standpoint of the mind, and, secondly, from that of the eye; after this has been done, we shall be better able to see whether these apparently antagonistic positions can be synthesized.

Though the visual stimuli transmitted to the brain have been, in the first place, received by the eye, it is continually brought home to us, in the affairs of everyday life, that whatever we see is seen *with the mind*.[1] We can often, for example, decipher a page of badly written manuscript without conscious effort, although, on focusing on the details of the script, it may be found that there is hardly a letter in it which is adequately formed, while many are even slurred over altogether. Yet from the retinal images of these distorted and imperfect symbols, the mind has leaped at once to an exactitude of elaborate meaning. Again, in looking at a landscape, we form a single mental picture, including

[1] For an account, by a psychologist, of vision as due to the co-operation of eye and brain, rather than to the eye alone, see Thouless, R. H. (1938).

foreground and remote background; but since the eye, as a mechanism, cannot receive clear impressions from different focal distances simultaneously, it can be only the mind which draws together this series of optical imprints, and fuses them into a unified whole.

We are too apt to think of pictorial images in the mind as if they had a quality of literal 'correspondent' truth to external objects; in actual fact they are based not merely upon the 'raw' retinal mosaic, but upon this mosaic as played upon and made into an image or pattern[1] by sensory organization. Such a pattern is a symbolic model, fashioned by the mind from materials supplied by the eye, rather than a photographic replica received passively by the mind.[2] Mental images of this kind have perhaps more in common with expressions of modern 'abstract' art, than with the familiar 'representational' art based on the 'copy' theory.[3] The mind, moreover, has the power, not only of modifying but of rejecting the data offered to it. When the whole mental attention is concentrated upon an individual object, we are scarcely conscious of 'seeing' anything else, though actually the sensory impression itself also includes the surroundings of whatever happens to be primary for us at the moment—surroundings which the mind deliberately ignores.

That we see *through* the eyes, rather than *with* them, is a point upon which Plato seems to have had little doubt, for he makes Socrates and Theaetetus agree that it is so.[4] The same idea is traceable in medieval literature. The philosopher-mystic, Meister Eckhart (d. *c.* 1327) said: "Subtract the mind,...and the eye is open to no purpose, which before did see"; and, again, "Before my eye can see the painting on the wall, this must...be borne into my phantasy, to be assimilated by my understanding."[5] Five centuries later, William Blake wrote: "I question not my Corporeal...Eye any more than I would Question a Window concerning a Sight. I look thro' it & not with it."[6]

[1] Russell, E. S. (1932), p. 168. [2] Rivaud, A. (1937), p. 297.
[3] For a study of this distinction, see Blanshard, F. B. (1949).
[4] Cornford, F. M. (1935), p. 103.
[5] Pfeiffer, F. (1924; 2nd imp., 1949); J. Eckhart's Tract. III, p. 288, and Serm. XLI, p. 111. (See note 2, Chapter I, p. 16, in the present book.)
[6] Blake, W., in Keynes, G. (1935), p. 135.

Every biologist must be able to confirm from his own experience that perception depends upon preparedness of mind, as well as on actual visual impressions. As a trivial instance, the writer may recall having been acquainted with Queen-Anne's-Lace (*Anthriscus sylvestris* Hofm.) for half a century, without noticing that the pattern of its growth is such that the main axis almost invariably terminates in a reduced inflorescence, which, in association with the grouping of the lateral shoots below it, gives the plant a highly distinctive facies.[1] When that visual fact had at last succeeded in forcing its way into the mind, any plant that came under observation was found to show this salient feature so strikingly as to leave the observer bewildered and humiliated at having been totally blind to it year after year.

How great a part the understanding plays, in comparison with the visual mechanism, when biological memoirs are being illustrated, comes home to one especially in looking at the simplified kind of figure which is so frequently used. In a black-and-white drawing, for example, it is evident that the visual impression actually received, which can have been only a mosaic of coloured patches, has been translated into a system of black marks on a white ground—marks which have no existence, as such, in Nature. This process is essentially symbolic and diagrammatic; it is an interpretation rather than a representation.

A symptom of the craving of the mind to mould sense impressions into a form consonant with itself, is the pleasure which mankind takes in symmetry.[2] This pleasure arises from the satisfaction of the characteristic mental urge to reduce the chaotic multifariousness of sense impressions to some unity of design. An urge of this type is illustrated by Coleridge's analogy of the kaleidoscope.[3] Into the toy, a medley of objects is introduced, which as a casual group can give the mind's eye no satisfaction—oddments such as bits of coloured glass, steel filings, scraps of silk, and cuttings of twine. The kaleidoscope transforms their meaningless jumble into a succession of harmonious phases;

[1] Arber, A. (1950), p. 185 and Fig. 42, p. 186.
[2] Cf. Lovejoy, A. O. (1948), p. 146.   [3] Snyder, A. D. (1929), p. 122.

as an instrument it thus creates a symmetry which has no existence in the actual congeries of objects. Examples of the free but disciplined type of symmetry that the mind imposes, may be recognized in certain landscapes by Chinese artists. The delicate and subtle sense of rhythm which these pictures reveal, is not derived simply from sense impressions, but is inspired by a mental awareness of the complete unity of man with the universe, which Western painters seldom realize so intensively.[1]

One may be fully convinced of the immense importance of the mind in seeing, and yet feel that many philosophers have stressed its dominance in a partisan fashion, which has led to the belittlement of the visual aspect of thought. This unbalanced attitude found extreme expression in the writings of Giordano Bruno, who said that the use of the senses is "solely to stimulate our reason", and that "truth is in but very small degree derived from the senses".[2] Kant, with characteristic insight, undermined such a standpoint as this in a famous passage. He wrote that "The light dove, cleaving the air as she freely wings her way, and feeling its resistance, might imagine that her flight would succeed a great deal better in airless space. In the same way, Plato, relinquishing the world of the senses, as setting such narrow limits to the understanding, ventured out beyond it on the wings of the ideas into the void space of the pure intellect. He did not perceive that he made no advance by all his efforts".[3] Few thinkers would regard this criticism as justly

---

[1] Binyon, L., in Binyon, L., and others (1935), p. 5; see also Richards, I. A. (1932) on the non-separation of human and external nature in Chinese thought.

[2] Bruno, G. (1923–7), vol. I, 1925, Dialogue I, *De l'Infinito Universo e Mondi*, p. 289: "Ad eccitar la raggione solamente,...Onde la verità, come da un debile principio, è da gli sensi in picciola parte, ma non è nelli sensi." For translation, see Singer, D. W. (1950), p. 251.

[3] Kant, I. (1902–38), Bd. III, 1904, *Kritik der Reinen Vernunft*, 2nd ed. 1787, *Einleitung*, III, p. 32: "Die leichte Taube, indem sie im freien Fluge die Luft theilt, deren Widerstand sie fühlt, könnte die Vorstellung fassen, dass es ihr im luftleeren Raum noch viel besser gelingen werde. Eben so verliess Plato die Sinnenwelt, weil sie dem Verstande so enge Schranken setzt, und wagte sich jenseit derselben auf den Flügeln der Ideen in den leeren Raum des reinen Verstandes. Er bemerkte nicht, dass er durch seine Bemühungen keinen Weg gewönne, denn er hatte keinen Widerhalt gleichsam zur Unterlage, worauf er sich steifen und woran er seine Kräfte anwenden konnte, um den Verstand von der Stelle zu bringen." Translation modified from Smith, N. Kemp (1933), p. 47, and Meiklejohn, J. M. D. (1934), p. 29.

applicable to Plato himself, but it is a valid comment on the speculations of many later writers. The biologist may perhaps venture to hope that it is by the infusion of the spirit of his own science into philosophy that the balance of the senses and the intellect may be redressed. Coleridge was an example of what Kant regarded as the Platonic extreme, when he wrote that it is essential for the achievement of abstract thought "To emancipate the mind from the despotism of the eye".[1] To speak of 'despotism', in this connexion, prejudices the case from the outset; it is an unfair word, since the eye is, rather, the servant of the mind, to which it offers all its data for interpretation; or it might perhaps be better to call the eye the 'junior partner' of the mind—the partner to whose vitality the firm owes much of its vigour. It is fortunate for the biologist that, whatever flight he may take into the empyrean, he is always bound to return to the solid ground of his own sensory data. All his instincts and principles incline him to an attitude towards the universe which does not discount *things* in favour of abstract *thought*; indeed an essential condition of worth-while biological work is the chastening of the mind through the discipline of the eye. So simple an example as the comparison of a well-stained section, seen under the microscope, with the best representation of it in black-and-white, reveals at once to the biologist how many subtleties, seizable by the eye, have been discarded from the intellectualized and impoverished drawing. One of the factors which have encouraged a certain depreciation of the visual element in intellectual processes, is the distinction drawn between 'primary' and 'secondary' qualities of bodies. This distinction is an ancient one, which was recognized by certain pre-Socratic philosophers;[2] among more recent writers, Locke[3] emphasized it strongly. He enumerates the "*primary Qualities*" as "Solidity, Extension, Motion or Rest, Number and Figure". On the other hand, he held that "*Colours* and *Smells*,... *Tastes*, and *Sounds*, *and other the like sensible Qualities*" are '*Secundary*', being "nothing

---

[1] Snyder, A. D. (1929), p. 126; see also Coleridge, S. T. (1817), vol. I, pp. 107–8.
[2] Burnet, J. (1914), pp. 196–7.
[3] Locke, J. (1690), bk. II, chap. VIII, pp. 56 *et seq.*

in the Objects themselves, but Powers to produce various Sensations in us", and that these powers depend upon the primary qualities. It was Berkeley[1] who detected the flaw in Locke's reasoning, and showed that the arguments, which led him to hold that secondary qualities exist only for the percipient, might be used equally well to prove that primary qualities, also, have no existence of their own. Although Locke's distinction between the two types of quality is no longer an accepted canon of the theory of knowledge, in residual form it still haunts philosophic thought, and is in part responsible for the unfortunate prejudice that allows only an inferior status to visual impressions.

Another factor which cramps the biologist's visual thinking is his tendency to share the general egoism which leads to concentration upon the human element in the universe, so that when he looks at an animal or plant he is liable to see it, not as it is in and for itself, but from an anthropocentric standpoint. The result is that the visual impression does not yield its full content, since it can offer this only to the mind that becomes one with what it sees, thus breaking down the rigid subject-object antithesis. Aquinas recognized this fact when he wrote that "Vision is made actual only when the thing seen is in a certain way in the seer".[2] Dante, St Thomas's younger contemporary, expressed the same conviction in the dictum, "Who paints the shape of ought, unless himself can be it, cannot set it down".[3] The oriental mind seems to 'feel itself' more naturally than the occidental into forms of life other than humanity; this is conspicuous in the work of Chinese and Japanese artists,[4] who often identify themselves, as it were, with a bird or a flower, thus revealing its individual character with an intuitive insight that

---

[1] Berkeley, G., in Jessop, T. E. (1949), vol. II, *The Principles of Human Knowledge*, pt. I, 15, p. 47; on Bayle's anticipation of Berkeley, see Luce, A. A. (1934), pp. 55, 64.

[2] Aquinas, St Thomas, in Migne, J. P. (1845, etc.), vol. I, p. 538, *Summa Theol.* I, quest. 12, art. 2, "non enim sit visio in actu nisi per quòd res visa quodammodò est in vidente". Translation in Aquinas, St Thomas (1911, etc.) I, no. 1, p. 122.

[3]   "Poi chi pinge figura
            Se non può esser lei, non la può porre."
Dante Alighieri. *Convivio*, Ode III, 3. Giuliani, G. (1874), p. 403.

[4] On this aspect of Chinese painting, see Binyon, L., in Binyon, L., and others (1935), p. x, and pp. 5–6; and Fry, R., in Fry, R., and others (1935), p. 4; see also p. 118 of the present chapter.

the European seldom achieves. In a modest way the biologist may occasionally feel a faint hint of such self-identification with the living thing which he is attempting to depict. An experience of this kind is indeed a precondition of any really happy portrayal, and it may be equated with that "faculté intuitive, qui ne se communique point, mais qui s'acquiert, jusqu'à un certain point, par la grande habitude de voir".[1] When this latent power is awakened, drawing becomes an intensified form of visual perception, in which the pencil, as well as the hand and the eye, is organically one with the brain. The artist's claim, that "learning to draw is learning to see", finds an echo in biological practice; it was a saying of Julius von Sachs to the students in his laboratory, that "Was man nicht gezeichnet hat, hat man nicht gesehen".[2] It is generally recognized that mathematicians are often gifted musically, while, from biographies of zoologists and botanists, one gathers the impression that biology has a corresponding relation to the visual arts. It might be interesting to subject this thesis to the test of large-scale statistical study.

It has been pointed out[3] that in European languages every word belongs to one of the 'parts of speech', and that we have no 'germinal' or 'ore-like'[4] terms, each of which can convey, in the form of a single symbol, a sentence not differentiated into nouns, verbs, etc., but enclosing, as it were 'in the bud', the synthetic significance of these elements. Indeed, the expression, 'parts of speech', itself involves the idea that 'speech' is primary, and that its analysis into words is merely secondary. When the biologist meets the difficulty of the lack of germinal words, he can sometimes circumvent it by transmitting his unanalysed meaning through illustrations. Artistic expression offers a mode of translation of sense data into thought, without subjecting them to the narrowing influence of an inadequate verbal framework; the verb, 'to illustrate', retains, in this sense, something of its ancient meaning—'to illuminate'.

---

[1] Turpin, P. J. F. (1820), p. 14.   [2] Scott, D. H. (1925), p. 10.
[3] Bosanquet, B. (1911), vol. I, p. 104.
[4] Cf. the use of 'ore-like' in Richards, I. A. (1932), p. 4.

An example of a modern movement of biological thought, in which the eye has come into its own, is the renaissance of botanical morphology which has taken place in twentieth-century Germany under the influence of Wilhelm Troll. The adherents of his school regard the concept of the flower (in the *Gestalt* sense), not as a totality of parts united according to a certain plan of organization, but as the totality of the flower itself, *as it appears to the eye*.[1] The eye, in this connexion, dominates discursive reasoning. The approach here is through Goethe's *Anschauung*, a term which has no equivalent in our language, but which may be interpreted as intuitive knowledge gained directly through contemplation of the visible aspect.[2]

Hitherto we have dwelt separately upon the two roads to reality open to the biologist—the way of the intellect and the way of the eye. We must now come to closer grips with the *relation* between these two approaches, in order to see if there is any possibility that, however distinct they may seem at a casual glance, they may prove, finally, to be one and indivisible.

It can scarcely be denied that the use of pictorial imagery in thinking is a fundamental need of the human mind. Though philosophers have often inveighed against it, as vitiating abstract thought, it cannot possibly be discarded from our mental tool-chest. Indeed, to employ it can do no harm, provided we never lose sight of what it is that we are about when we handle it. A statue of his god may be helpful to a worshipper who sees it as a symbol. There is, however, the risk that he will take it at its face value, when it becomes an idol; for, if we may set a construction other than his own upon the words that Plato puts into the mouth of Socrates; "There are few who, going to the images, behold in them the realities."[3] Our whole language is, moreover, permeated with metaphor, but this is misleading only if we forget that such pictorial expressions are merely symbolic.

[1] Troll, W. (1928), p. 88; on *Gestalt* morphology, see Arber, A. (1950), pp. 144–61.
[2] Cf. Arber, A. (1950), pp. 209–11; see also Arber, A. (1946a), p. 85.
[3] Jowett, B. (1871), vol. I, *Phaedrus*, 250, p. 584.

Kant[1] pointed to a bridge between abstract and sensuous thinking, when he showed how images representing percepts could be replaced in abstract thought by 'schemata' of concepts. These are mental portrayals of the Platonic 'forms' of objects, rather than of individual objects. The *schema* of the triangle exists in thought alone, and is adequate universally for triangles, whereas the *image* of a triangle can never attain to this universality.

The more one considers the matter, the clearer it becomes that sense perceptions cannot be separated from mental activity.[2] This was recognized in the seventeenth century by Descartes, who—according to one of his modern interpreters—regarded *l'entendement* (conceptual thinking) and *l'imagination* (pictorial thinking) as indissolubly united in perception, of which conceptual thinking is the form, and picture thinking, the matter.[3] The mind, indeed, at once seizes on the material which the visual apparatus offers to it, and forms a 'perceptual judgment',[4] in which the sense impression and the interpretative work of the intellect are inextricably blended. This union recalls Hegel's conception of *reason* as the synthesis of intuitive sensation and discursive thought,[5] and Bosanquet's "reason in the form of feeling".[6] Walter Pater, again, approaching the subject rather from the standpoint of literature and art than from that of philosophy, called the same faculty the imaginative reason, "for which every thought and feeling is twin-born with its sensible analogue or symbol".[7]

The fallacious notion of a sharp disjunction between mental and visual thinking, has its physical basis in the conception of brain and sense organs as discrete entities; but there is no such

[1] Kant, I. (1902–38), vol. III, 1904, *Kritik der reinen Vernunft*, 2nd ed., 1787, Elementarlehre, Th. II, Abth. I, Buch 2, Hauptst. I, pp. 133–9, *Von der Schematismus der reinen Verstandesbegriffe*. Translation in Smith, N. Kemp (1933), pp. 180–7, and Meiklejohn, J. M. D. (1934), pp. 117–22.
[2] Cf. Joachim, H. H. (1948), p. 83.
[3] Lévêque, R. (1923), p. 24; Lévêque bases his view on Descartes, R. (1641), Med. II, pp. 25 *et seq.* For translation, see Haldane, E. S., and Ross, G. R. T. (1911–12), vol I, pp. 154 *et seq.*
[4] Johnson, W. E. (1921–4), pt. I, pp. xvi, xvii.
[5] Walsh, W. H. (1947), pp. 27, 28.
[6] Bosanquet, B. (1911), vol. II, p. 236.
[7] Pater, W. (1888), *The School of Giorgione*, p. 144.

severance in reality, though there is differentiation and division of labour. The whole nervous system, including the brain, is a unity though an inexpressibly complex one. This is true on the plane of 'extension', and equally so on the plane of 'thought'; the activities of the sense organs, and the thinking of the brain, are all parts of an indivisible whole. When Donne, writing at the end of the sixteenth century, not as a physiologist but as a poet, said:

> For if the sinewie thred my braine lets fall
> Through every part,
> Can tye those parts, and make me one of all;[1]

he was recognizing intuitively that the heterogeneous complexities of the body are unified by the brain and nervous system.

The close interlocking and interweaving of the data gained directly through the senses, with the concepts of pure thought, are peculiarly marked in the biologist's sphere of work. He may with conviction echo Kant's words: "The understanding can intuit nothing, the senses can think nothing. Only through their union can knowledge arise."[2] At an even earlier date, Diderot, in his *Thoughts on the Interpretation of Nature*, had stressed the continual to-and-fro movement between the senses and reflective thinking. "Tout se réduit", he wrote, "à revenir des sens à la réflexion, et de la réflexion aux sens; rentrer en soi et en sortir sans cesse...."[3]

Some of those workers for whom the physico-chemical disciplines occupy the centre of the stage, define science in a way that excludes any other approach than that used in these subjects in which visual thinking is at a discount. It has been said, for instance, that "the making of science, and science itself, as a living enterprise, is inductive", and it has been suggested that its concern is "the prediction of the unperceived from the perceived".[4] Though this may be true of physics and chemistry,

---

[1] [Donne, J.] 'J.D.' (1635), *The Funerall*, p. 51.
[2] Kant, I. (1902–38), vol. III, 1904, *Kritik der reinen Vernunft*, 2nd ed. 1787, *Elementarlehre*, Th. 2, *Trans. Log.*, *Einleitung*, I, p. 75: "Der Verstand vermag nichts anzuschauen und die Sinne nichts zu denken. Nur daraus, dass sie sich vereinigen, kann Erkenntniss entspringen." Translation in Smith, N. Kemp (1933), p. 93.
[3] Diderot, D. (1875–7). *Pensées sur l'interprétation de la nature* (1754), vol. II, p. 14 (ix).
[4] Williams, D. (1947), pp. 12–13.

and of those phases of biological study in which physico-chemical methods predominate, it does not, for example, indicate adequately the goal of pure morphology, which might be described as *the visual and conceptual interpretation of the perceived,* rather than as *the conceptual prediction of the unperceived.*

The morphological approach to biology is through structure viewed as form. Structure is a relational category, which may be defined as the arrangement or organization of parts within an integrated whole.[1] Problems of pure morphology cannot be solved by the methods of analytical science. The contemplative treatment of comparative form, rather than its analysis from the standpoint of cause and effect, becomes the morphologist's aim; he desires to see form, both with the bodily eye and with the mind's eye, not only in itself, but in its nexus of relations. This process of mental visualization differs essentially from the thought-techniques of the physico-chemical disciplines. The morphologist's standpoint is set midway between that of the mechanistic sciences and of the arts, so that his work should offer a synthesis of intellectualist logic and sensory apprehension. In other words, he brings into use Aristotle's conception of the objects of our senses as 'immattered form', through which the contrast is bridged between "pure thought and the empirical study of individuals".[2] In this respect, the view point of Aristotle—the Father of Biology—combines that of Democritus, for whom sensation was primary, and that of Plato, who laid his chief stress upon conceptual thinking.[3] There is much to be said for the suggestion that, whereas *Metaphysics* studies 'being' as such, and *Natural Science* (of the physico-chemical type) treats of the corporeal world, *Natural Philosophy* may be so defined as to link the two; it would then connote that mental activity which ceaselessly weaves connexions between the planes of intangible 'essence' and tangible 'existence'.[4] Long ago Coleridge realized that natural science, though beginning with material

[1] Wood, L. (1940), p. 180.   [2] Jaeger, W. (1948) cf. pp. 340–1.
[3] Aquinas, St Thomas, in Migne, J. P. (1845, etc.), vol. I, pp. 1162–3, *Summa Theol.* I, quest. 84, art. 6. Translation, Aquinas, St Thomas (1911, etc.), I, no. 3, p. 170 *et seq.*; see also Hawkins, D. J. B. (1947), p. 78.
[4] This idea is a version of that expressed by Thompson, W. R. (1937), pp. 131–2.

phenomena, is transformed, finally, into natural philosophy.[1]
It is to this synthetic discipline that biology, in its autonomous
aspect, belongs;[2] its function is fulfilled when it offers its mite
towards the ultimate fusion of metaphysical and scientific
thinking.

[1] Coleridge, S. T. (1817), vol. I, p. 258.
[2] See also Chapter VI, p. 71. After the first draft of the present book was
completed, Beveridge, W. I. B. (1950), came into the writer's hands. In Beveridge's
work some of the topics, discussed in the preceding pages, will be found treated
from a contrasting standpoint, since biology, as he understands it, is a branch of
pure Natural Science, and essentially experimental, rather than a part of what has
been distinguished here as Natural Philosophy.

LIST OF BOOKS AND MEMOIRS CITED

WITH PAGE REFERENCES TO THE TEXT OF THE PRESENT BOOK

ADAM, C.; see DESCARTES, R. (1897–1910).

ADLER, M. J. (1927). *Dialectic*. London. [Pp. 45, 94, 113.]

AINSLIE, D.; see CROCE, B. (1915).

ALBERTUS MAGNUS; see DIGBY, Sir K. (1654).

ANDRADE, E. N. DA C. (1947). Newton, *Royal Society, Newton Tercentenary Celebrations*. Cambridge. Pp. 3–23. [P. 29.]

AQUINAS, ST THOMAS, in Migne, J. P. (1845, etc.), *Petrologiae cursus completus*. Ser. sec. [Pp. 24, 64, 120, 125.]

AQUINAS, ST THOMAS (1911, etc.). *Summa Theologica*. Trans. by Fathers of the English Dominican Province. London. [Pp. 24, 64, 120, 125.]

ARBER, A. (1920). *Water Plants: a study of Aquatic Angiosperms*. Cambridge. [P. 102.]

ARBER, A. (1934). *The Gramineae: a study of Cereal, Bamboo, and Grass*. Cambridge. [P. 102.]

ARBER, A. (1937). The interpretation of the flower: a study of some aspects of morphological thought. *Biol. Rev.* vol. XII, pp. 157–84. [P. 43.]

ARBER, A. (1941 *a*). Tercentary of Nehemiah Grew (1641–1712). *Nature*, vol. CXLVII, pp. 630–2, May 24. [P. 42.]

ARBER, A. (1941 *b*). The relation of Nehemiah Grew and Marcello Malpighi. *Chron. Bot.* vol. VI, pp. 391–2. [P. 42.]

ARBER, A. (1941 *c*). Nehemiah Grew and Marcello Malpighi. *Proc. Linn. Soc. Lond.* Session 153, pt. 2, Nov. 21, pp. 218–38. [P. 42.]

ARBER, A. (1942). Nehemiah Grew (1641–1712) and Marcello Malpighi (1628–1694): an essay in comparison. *Isis*, vol. XXXIV, pp. 7–16. [P. 42.]

ARBER, A. (1946 *a*). Goethe's Botany. *Chronica Botanica* (Waltham, Mass. U.S.A.), vol. X, no. 2, pp. 63–126 (separate publication). [Pp. 106, 122.]

ARBER, A. (1946 *b*). Analogy in the history of science, pp. 221–33 in *Studies and Essays in the History of Science and Learning*, offered to George Sarton, Ed. by M. F. Ashley Montagu. New York. [P. 32.]

ARBER, A. (1950). *The Natural Philosophy of Plant Form*. Cambridge. [Pp. 28, 41, 43, 81, 93, 100, 101, 102, 105, 109, 117, 122.]

ARISTOTLE (translations); see BYWATER, I. (1909), MURE, G. R. G. (1926), TREDENNICK, H. (1933).

AUBREY, J. (1950). *Aubrey's Brief Lives*, edited from the original manuscripts by O. L. Dick. London. [P. 51.]

BACON, F. (1620). *Instauratio magna*. London. [Pp. 27, 32, 33.]

BACON, F.; see KITCHIN, G. W. (1855).

BAILLIE, J. B.; *see* HEGEL, G. W. F. (1931).

BEER, G. R. DE; *see* YOUNG, J. Z. (1938*a*).

BERKELEY, G.; *see* JESSOP, T. E. (1949).

BERTALANFFY, L. VON (1933). *Modern Theories of Development: an Introduction to Theoretical Biology*. Trans. by J. H. Woodger. Oxford. [P. 101.]

BEVERIDGE, W. I. B. (1950). *The Art of Scientific Investigation*. London. [Pp. 47, 126.]

BINYON, L., and others (1935). *Chinese Art*. London. [Pp. 118, 120.]

BLAKE, W.; *see* KEYNES, G. (1925) *and* (1935).

BLANSHARD, F. B. (1949). *Retreat from Likeness in the Theory of Painting*. 2nd ed. New York. [Pp. 45, 116.]

BOSANQUET, B. (1911). *Logic or the Morphology of Knowledge*. 2 vols. (1st ed. 1888). Oxford. [Pp. 22, 46, 59, 70, 82, 93, 121, 123.]

BOSANQUET, B. (1920). *Implication and Linear Inference*. London. [Pp. 23, 46, 67, 87.]

BOSANQUET, B.; *see* MUIRHEAD, J. H. (1935).

BOUTROUX, E.; *see* LEIBNIZ, G. W. (1930).

BOYCOTT, A. E. (1929). The transition from live to dead: the nature of filtrable viruses. *Nature*, vol. CXXIII, Jan. 19, pp. 93–8. [P. 6.]

BOYLE, THE HON. R. (1680). *The Sceptical Chymist*. Oxford. [Pp. 28, 112.]

BRADFIELD, J. R. G. (1950). Biochemical aspects of cell morphology (in Discussion on morphology and fine structure). *Proc. Linn. Soc. Lond.* vol. CLXII, pt. I, pp. 76–81. [Pp. 9, 94.]

BRADLEY, F. H. (1914). *Essays on Truth and Reality*. Oxford. [P. 96]

BRADLEY, F. H. (1922). *The Principles of Logic*. 2 vols. 2nd ed. Oxford. [P. 72.]

BRADLEY, F. H. (1930). *Appearance and Reality*. (1st ed. 1893.) [P. 72.]

BRÉHIER, E. (1924–38). *Plotin. Ennéades*. Paris. [P. 91.]

BRODRICK, J. (1928). *The Life and Work of Cardinal Bellarmine (1542–1621)*. 2 vols. London. [P. 24.]

BROWN, T. K.; *see* NELSON, L. (1949).

BROWNE, Sir T. (1928–31). *Works*. Ed. by G. Keynes. 6 vols. (The *Religio Medici*, written in 1635, is reproduced from the edition of 1682.) London. [Pp. 39, 40.]

BRUNO, G. (1923–7). *Opere Italiane*. Ed. by B. Croce and V. Spampanato. 3 vols. Bari. [Pp. 106, 110, 111, 118.]

BRUNO, G.; *see* GREENBERG, S. (1950) *and* SINGER, D. W. (1950).

BURNET, J. (1914). *Greek Philosophy*, part I, *Thales to Plato*. London. [Pp. 24, 77, 119.]

BURNET, J. (1920). *Early Greek Philosophy*. 3rd ed. [Pp. 12, 106.]

BUTLER, J. (1736). *The Analogy of Religion, Natural and Revealed, to the Constitution and Course of Nature*. London. [Pp. 33, 36.]

BYWATER, I. (1877). *Heracliti Ephesii Reliquiae*. Oxford. [P. 12.]

BYWATER, I. (1909). *Aristotle on the Art of Poetry*. Text with translation. Oxford. [P. 32.]

CAIRD, E. (1889). *The Critical Philosophy of Immanuel Kant*. 2 vols. Glasgow. [Pp. 10, 16, 78, 104, 109.]

CAMPBELL, N. R. (1920). *Physics: the Elements*. Cambridge. [P. 58.]

CANDOLLE, A. P. DE (1813). *Théorie Élémentaire de la Botanique.* Paris. [P. 93.]

CARR, H. WILDON (1930). *The Monadology of Leibniz.* Trans. with Introd., Commentary, etc. London. [P. 39.]

CARRÉ, M. H. (1946). *Realists and Nominalists.* Oxford. [P. 88.]

CARRÉ, M. H. (1949). *Phases of Thought in England.* Oxford. [P. 27.]

CARRÉ, M. H.; see HERBERT, E. (Lord Herbert of Cherbury) (1937).

CLARK, J. M. (1949). *The Great German Mystics, Eckhart, Tauler and Suso* (Modern Language Studies, v). Oxford. [P. 16.]

CLARKE, J., and GEIKIE, Sir A. (1910). *Physical Science in the Time of Nero, being a translation of the 'Quaestiones Naturales' of Seneca.* London. [P. 38.]

CLIFFORD, W. K. (1879). On the aims and instruments of scientific thought. Delivered at Brit. Ass. Adv. Sci. Brighton, 1872. Published in *Lectures and Essays.* Ed. by L. Stephen and F. Pollock, vol. 1, pp. 124–57 [P. 83.]

COBURN, K. (1951). *Inquiring Spirit: a new Presentation of Coleridge from his... Prose Writings.* London. [P. 98.]

COLERIDGE, E. H.; see COLERIDGE, S. T. (1895).

COLERIDGE, S. T. (1817). *Biographia Literaria.* 2 vols. London. [Pp. 10, 49, 98, 119, 126.]

COLERIDGE, S. T. (1818). *The Friend.* New ed. vol. 1. London. [P. 96.]

COLERIDGE, S. T. (1895). *Anima Poetae.* From the unpublished note-books of S. T. C., edited by E. H. Coleridge. London. [P. 105.]

COLERIDGE, S. T.; see COBURN, K. (1951) *and* SNYDER, A. D. (1929).

COLLINGWOOD, R. G. (1924). *Speculum Mentis or the Map of Knowledge.* Oxford. [P. 58.]

COLLINGWOOD, R. G.; see DE RUGGIERO, G. (1921).

COOMARASWAMY, A. K. (1934). *The Transformation of Nature in Art.* Harvard Univ. Press, Cambridge, Mass. [P. 68.]

CORNFORD, F. M. (1932). *Before and After Socrates.* Cambridge. [P. 54.]

CORNFORD, F. M. (1935). *Plato's Theory of Knowledge. The 'Theaetetus' and the 'Sophist' of Plato translated with a running commentary.* London. [Pp. 25, 35, 116.]

CORNFORD, F. M. (1937). *Plato's Cosmology. The 'Timaeus' translated with a running commentary.* London. [Pp. 33, 38, 47, 69, 82.]

CORNFORD, F. M. (1939). *Plato and Parmenides. Parmenides' 'Way of Truth' and Plato's 'Parmenides' translated with an introduction and a running commentary.* London. [P. 103.]

COUTURAT, L. (1901). *La Logique de Leibniz.* Paris. [P. 90.]

CROCE, B. (1915). *What is Living and What is Dead of the Philosophy of Hegel.* Trans. from the 3rd Italian ed. by D. Ainslie. London. [P. 97.]

CROCE, B.; see BRUNO, G. (1923–7).

CURTIS, S. J. (1950). *A Short History of Western Philosophy in the Middle Ages.* London. [P. 88.]

DAICHES, D. (1940). *Poetry and the Modern World.* Chicago. [P. 44.]

DANTE ALIGHIERI (1900). *La Commedia...il testo Wittiano riveduto da Paget Toynbee.* London. [P. 83.]

DANTE ALIGHIERI (1916). *De Monarchia.* Oxford text ed. by E. Moore. Oxford. [P. 88.]

DANTE ALIGHIERI; see GIULIANI, G. (1874) *and* JACKSON, W. W. (1909).

DARWIN, F. (1888). *The Life and Letters of Charles Darwin.* Ed. by F. Darwin. 3 vols. London. [P. 66.)

DAUDIN, H. (1926). *Les Classes zoologiques et l'idée de série animale en France à l'époque de Lamarck et de Cuvier (1790–1830).*[1] 2 vols. Paris. [P. 104.]

DAVIS, W.; see NEWTON, Sir I. (1803).

DE BEER, G. R.; see YOUNG, J. Z. (1938a).

DE MORGAN, A. (1847). *Formal Logic.* London. [P. 108.]

DE RUGGIERO, G. (1921). *Modern Philosophy.* Trans. by A. H. Hannay and R. G. Collingwood. London. [P. 69.]

[DESCARTES, R.] (1637). *Discours de la methode.* Leyden. [Pp. 6, 7, 45, 98.]

DESCARTES, R. (1641). *Meditationes de prima philosophia.* Paris. [Pp. 98, 99, 123.]

DESCARTES, R. (1644). *Principia Philosophiae.* Amsterdam. [P. 98.]

DESCARTES, R. (1897–1910). *Œuvres.* Ed. by C. Adam and P. Tannery. 12 vols. Paris. [P. 111.]

DESCARTES, R. (1947). *Discours de la méthode.* Texte et commentaire par É. Gilson. Paris. [Pp. 6, 7, 45, 98.]

DESCARTES, R. (translation); see HALDANE, E. S., and Ross, G. R. T. (1911–12).

DICK, O. L.; see AUBREY, J. (1950).

DIDEROT, D. (1875–7). *Œuvres complètes* (20 vols.). *Pensées sur l'interprétation de la nature* (1754), vol. II, 1875, pp. 9–60. Paris. [P. 124.]

DIGBY, Sir K. (1654). *A Treatise of Adhering to God; Written by Albert the Great, Bishop of Ratisbon.*[2] *Put into English by Sir Kenelme Digby, Kt.* London. [Pp. 53, 54.]

DINGLE, H. (1931). *Science and Human Experience.* London. [P. 25.]

DINGLE, H. (1937). *Through Science to Philosophy.* Oxford. [P. 36.]

DOBELL, C. (1951). In Memoriam Otto Bütschli (1848–1920), 'Architect of Protozoology'. *Isis,* vol. XLII, pp. 20–2. [P. 58.]

DODDS, E. R. (1923). *Select Passages Illustrating Neoplatonism.* London. [Pp. 29, 74.]

DODDS, E. R. (1933). *Proclus. The Elements of Theology. A Revised Text with Translation, Introduction and Commentary.* Oxford. [P. 91.]

[DONNE, J.] 'J.D.' (1635). *Poems.* London. [P. 124.]

DORMER, K. J. (1948). Some geometrical considerations regarding imbricate aestivation. *Nature,* vol. CLXII, pp. 653–4. [P. 108.]

ECKHART, J.; see PFEIFFER, F. (1924) *or* (1949).

EINSTEIN, A. (1940). Science and religion. *Nature,* vol. CXLVI, Nov. 9, pp. 605–7. [P. 85.]

EINSTEIN, A.; see NEWTON, Sir I. (1931).

ELIOT, G. (1871–2). *Middlemarch.* Edinburgh and London. [P. 97.]

EVANS, C. DE B.; see PFEIFFER, F. (1924) *or* (1949).

EYRE, L. B.; see NORDENSKIÖLD, N. E. (1950).

FARBER, E. (1950). Chemical discoveries by means of analogies. *Isis,* vol. XLI, pp. 20–6. [P. 32.]

[1] It seems impossible to determine which of various titles the author intended his book to bear.

[2] Not now attributed to Albertus Magnus.

FISCHER, K. (1857). *Francis Bacon of Verulam.* Trans. by J. Oxenford. London. [P. 32.]

FLUGEL, J. C. (1933). *A Hundred Years of Psychology, 1833–1933.* London. [P. 7.]

FRISCH, C.; *see* KEPLER, J. (1858–70).

FRY, R., and others (1935). *Chinese Art.* London. [Pp. 11, 120.]

FUNG YU-LAN (1947). *The Spirit of Chinese Philosophy.* Trans. by E. R. Hughes. London. [P. 20.]

GALTON, F. (1889). *Natural Inheritance.* London. [Pp. 40, 41.]

GEIKIE, A.; *see* CLARKE, J., and GEIKIE, Sir A. (1910).

GEOFFROY SAINT-HILAIRE, É. (1830). *Principes de philosophie zoologiques, discutés en mars 1830, au sein de l'Académie royale des sciences.* Paris. [P. 104.]

GERHARDT, K. I.; *see* LEIBNIZ, G. W. (1875–90).

GIBBON, E. (1896). *The Autobiographies.* Ed. by John Murray. London. [P. 11.]

GILSON, É.; *see* DESCARTES, R. (1947).

GIULIANI, G. (1874). *Il Convito di Dante Alighieri.* Firenze. [Pp. 77, 120.]

GREENBERG, S. (1950). *The Infinite in Giordano Bruno. With a translation of his dialogue, Concerning the Cause, Principle, and One.* Columbia University, New York. [P. 111.]

GREGORY, J. C. (1945). On knowing one another. *Philosophy,* vol. xx, pp. 244–55. [P. 32.]

GREW, N. (1673). *An Idea of a Phytological History...Particularly prosecuted upon Roots.* London. [P. 34.]

GREW, N. (1681). *The Comparative Anatomy of Stomachs and Guts Begun.* Printed with *Musaeum Regalis Societatis.* London. [P. 34.]

GREW, N. (1682). *The Anatomy of Plants. With an Idea of a Philosophical History of Plants.* London. [P. 42.]

GUÉNON, R. (1945). *Introduction to the Study of the Hindu Doctrines.* Trans. by M. Pallis. London. [P. 106.]

HADAMARD, J. (1945). *An Essay on The Psychology of Invention in the Mathematical Field.* Princeton. [P. 18.]

HALDANE, E. S., and ROSS, G. R. T. (1911–12). *The Philosophical Works of Descartes.* Translation. 2 vols. Cambridge. [Pp. 6, 7, 45, 98, 99, 111, 123.]

HALDANE, J. S. (1935). *The Philosophy of a Biologist.* Oxford. [P. 103.]

HALLER, A. VON (1768). *Historia stirpium indigenarum Helvetiae.* 2 vols. Bern. [P. 46.]

HAMILTON, Sir W. (1852). *Discussions.* London. [P. 88.]

HANNAY, A. H.; *see* DE RUGGIERO, G. (1921).

HARDING, R. E. M. (1948). *An Anatomy of Inspiration.* 3rd ed. Cambridge. [P. 17.]

HARRIS, C. R. S. (1927). *Duns Scotus.* 2 vols. Oxford. [P. 88.]

HARVEY, J. W.; *see* METZ, R. (1938; reimpression, 1950).

HATFIELD, H. S.; *see* MONTMASSON, J.-M. (1931).

HAWKINS, D. J. B. (1947). *A Sketch of Mediaeval Philosophy.* London. [P. 125.]

HEGEL, G. W. F. (1931). *The Phenomenology of Mind.* Trans. with introd. by J. B. Baillie. London. [Pp. 94, 109.]

HEGEL, G. W. F.; *see* WALLACE, W. (1874).

HERACLITUS (HERAKLEITOS); *see* BYWATER, I. (1877) *and* PATRICK, G. T. W. (1888), (1889).

HERBERT, E. (Lord Herbert of Cherbury) (1937). *De Veritate*. Trans. with Introd. by M. H. Carré. Bristol. [Pp. 72, 74, 86.]

HERSCHEL, J. F. W. (1831). *Preliminary Discourse on the Study of Natural Philosophy*. The Cabinet Cyclopaedia (D. Lardner). (Date on engraved title-page, 1830.) London. [P. 89.]

HIBBEN, J. G. (1910). *The Philosophy of the Enlightenment*. London. [P. 87.]

HILDEGARD OF BINGEN; *see* SINGER, C. (1917).

HIRIYANNA, M. (1949). *The Essentials of Indian Philosophy*. London. [P. 106.]

HOBBES, T., of Malmesbury (1651). *Leviathan*. London. [Pp. 67, 72.]

HOBHOUSE, L. T. (1896). *The Theory of Knowledge*. London. [Pp. 23, 70.]

HUGHES, E. R.; *see* FUNG YU-LAN (1947).

HULTSCH, F. (1876–8). *Pappi Alexandrini collectionis quae supersunt*. 3 vols. Berlin. [P. 77.]

HUME, D. (1854). *The Philosophical Works*. 4 vols. Boston and Edinburgh. [Pp. 87, 112.]

HUME, D.; *see* SMITH, N. KEMP (1947).

HUXLEY, T. H.; *see* OWEN, R. (1894).

JACKSON, W. W. (1909). *Dante's 'Convivio' translated into English*. Oxford. [P. 77.]

JAEGER, W. (1947). *The Theology of the Early Greek Philosophers*. Gifford Lectures, 1936. Oxford. [P. 13.]

JAEGER, W. (1948). *Aristotle*. Trans. by R. Robinson. 2nd ed. Oxford. [P. 125.]

JEFFREYS, Sir H. (1937). *Scientific Inference*. Reissued with addenda. Cambridge. [P. 89.]

JESSOP, T. E. (1949). Vol. II of *Works of George Berkeley Bishop of Cloyne*, ed. by A. A. Luce and T. E. Jessop. Includes *A Treatise concerning the Principles of Human Knowledge* (1710), pp. 1–113, and *Three Dialogues between Hylas and Philonous* (1713), pp. 147–263. London and Edinburgh. [Pp. 33, 112, 120.]

JESSOP, T. E.; *see* METZ, R. (1938; reimpression, 1950).

JEVONS, W. S. (1877). *The Principles of Science*. 2nd ed. London. [P. 33.]

JOACHIM, H. H. (1901). *A Study of the Ethics of Spinoza*. Oxford. [P. 84.]

JOACHIM, H. H. (1939). *The Nature of Truth*. 2nd ed. (1st ed. 1906). Oxford. [Pp. 67, 70.]

JOACHIM, H. H. (1948). *Logical Studies*. Oxford. [Pp. 22, 59, 106, 123.]

JOHNSON, S. (1775). *A Journey to the Western Islands of Scotland*. Dublin. [P. 51.]

JOHNSON, W. E. (1921–4). *Logic*. Parts I, II, III. Cambridge. [Pp. 22, 23, 31, 65, 123.]

JOWETT, B. (1871). *The Dialogues of Plato*. Translation. Oxford. (*or see* 3rd ed. 1892, reimp. 1931) [Pp. 49, 77, 122.]

KABITZ, W. (1909). *Die Philosophie des jungen Leibniz*. Heidelberg. [P. 15.]

KANT, I. (1783). *Prolegomena zu einer jeden künstigen Metaphysik*. Riga. [P. 78.]

KANT, I. (1902–38). *Gesammelte Schriften*. Herausgegeben v.d.k. Preuss. Akad. d. Wiss. Berlin. [Pp. 10, 22, 66, 78, 106, 109, 118, 123, 124.]

KANT, I.; *see* KEHRBACH, K. (1919), MEIKLEJOHN, J. M. D. (1934) *and* SMITH, N. KEMP (1923), (1933).

KEHRBACH, K. (1919). *I. Kant. Kritik der reinen Vernunft. Text der Ausgabe von 1781.* Leipzig. [P. 85.]

KEPLER, J. (1858–70). *Opera Omnia.* Ed. by C. Frisch. 8 vols. Frankfort. (*Harmonices Mundi,* 1619, vol. v, 1864, pp. 75–327.) [P. 39.]

KEYNES, G. (1925). *The Writings of William Blake.* 3 vols. London. [P. 74.]

KEYNES, G. (1935). *The Note-book of William Blake called the Rossetti manuscript.* London. [Pp. 49, 116.]

KEYNES, G.; *see* BROWNE, Sir T. (1928–31).

KEYNES, J. M. (Lord) (1921). *A Treatise on Probability.* London. [P. 26.]

KEYNES, J. M. (Lord) (1947). Newton, the man. *Royal Society, Newton Tercentenary Celebrations.* Pp. 27–34. Cambridge. [P. 47.]

KITCHIN, G. W. (1855). *F. Bacon. The Novum Organon.* Translation. Oxford. [Pp. 27, 32.]

KNEALE, W. (1949). *Probability and Induction.* Oxford. [Pp. 26, 27, 83.]

KONRAD VON MEGENBERG; *see* PFEIFFER, F. (1861)

LAND, J. P. N.; *see* VLOTEN, J. VAN, and LAND, J. P. N. (1882–3).

LEIBNIZ, G. W. (1875–90). *Die philosophischen Schriften.* Ed. by K. I. Gerhardt. 7 vols. Berlin. [P. 83.]

LEIBNIZ, G. W. (1930). *La Monadologie publiée d'après les manuscrits.* E. Boutroux and H. Poincaré. 13th ed. Paris. [P. 100.]

LEIBNIZ, G. W.; *see* CARR, H. WILDON (1930).

LÉVÊQUE, R. (1923). *La Problème de la vérité dans la philosophie de Spinoza.* Strasbourg. [P. 123.]

LOCKE, J. (1690). *An Essay concerning Humane Understanding.* London. [Pp. 10, 119.]

LOEWENBERG, K.; *see* MEYERSON, É. (1930).

LOVEJOY, A. O. (1948). *Essays in the History of Ideas.* Baltimore. [P. 117.]

LUCE, A. A. (1934). *Berkeley and Malebranche.* Oxford. [P. 120.]

LUCE, A. A.; *see* JESSOP, T. E. (1949).

MACKENZIE, J. S. (1917). *Elements of Constructive Philosophy.* London. [P. 22.]

McTAGGART, J. McT. E. (1922). *Studies in the Hegelian Dialectic.* (1st ed. 1896.) Cambridge. [Pp. 48, 81, 110.]

MASSON, D. (1874). *The Poetical Works of John Milton.* Ed. London. [P. 11.]

MEGENBERG, K. VON; *see* PFEIFFER, F. (1861).

MEIKLEJOHN, J. M. D. (1934). *Kant's Critique of Pure Reason.* Translation. Everyman's Library. London. [Pp. 118, 123.]

METZ, R. (1938; reimpression, 1950). *A Hundred Years of British Philosophy.* Trans. by J. W. Harvey, T. E. Jessop and H. Sturt. Ed. by J. H. Muirhead. London. [Pp. 23, 70, 74.]

MEYER, A. (1900). Wesen und Geschichte der Theorie von Mikro- und Makrokosmos. *Berner Studien zu Philosophie und ihrer Geschichte,* vol. xxv, pp. 1–122. [Pp. 36, 37.]

MEYERSON, É. (1930). *Identity and Reality.* Trans. by K. Loewenberg. London. [P. 17.]

MEYERSON, É. (1931). *Du cheminement de la pensée.* 3 vols. Paris. [Pp. 3, 81, 103.]

MIGNE, J. P.; *see* AQUINAS, ST THOMAS (1845, etc.).

MILL, J. S. (1843). *A System of Logic.* 2 vols. London. [Pp. 27, 41, 80, 83, 114.]

MILTON, J.; *see* MASSON, D. (1874).

MONTAGU, M. F. ASHLEY; *see* ARBER, A. (1946 *b*).

MONTMASSON, J.-M. (1931). *Invention and the Unconscious.* Trans. by H. S. Hatfield. London. [P. 17.]

MOORE, E.; *see* DANTE ALIGHIERI (1916).

MORGAN, A. DE; *see* DE MORGAN, A. (1847).

MOTTE, A.; *see* NEWTON, Sir I. (1803).

MUIRHEAD, J. H. (1930). *Coleridge as Philosopher.* London. [P. 57.]

MUIRHEAD, J. H. (1931; reimpression, 1939). *The Platonic Tradition in Anglo-Saxon Philosophy.* London. [Pp. 81, 106, 107.]

MUIRHEAD, J. H. (1935). *Bernard Bosanquet and his Friends.* London. [P. 72.]

MUIRHEAD, J. H.; *see* METZ, R. (1938; reimpression, 1950).

MURE, G. R. G. (1926). *Analytica Posteriora* (in *The Works of Aristotle Translated into English,* vol. II). Oxford. [P. 77.]

MURE, G. R. G. (1932). *Aristotle.* London. [Pp. 94, 101, 109, 112.]

MURE, G. R. G. (1940). *An Introduction to Hegel.* Oxford. [P. 70.]

MURRAY, J.; *see* GIBBON, E. (1896).

NELSON, L. (1949). *Socratic Method and Critical Philosophy.* Trans. by T. K. Brown. Yale University Press, New Haven, U.S.A. [P. 47.]

NEWMAN, J. H. (1870). *An Essay in aid of a Grammar of Assent.* London. [P. 74.]

NEWTON, Sir I. (1803). *The Mathematical Principles of Natural Philosophy.* Trans. by Andrew Motte. New ed. revised by W. Davis. 3 vols. London. [Pp. 57, 89.]

NEWTON, Sir I. (1931). *Opticks.* Reprint of 4th ed., 1730. Foreword by A. Einstein and Introduction by Sir E. T. Whittaker. London. [Pp. 20, 26, 29, 83.]

NORDENSKIÖLD, N. E. (1950); (1st ed. 1928). *The History of Biology.* Trans. from the Swedish (1920–4) by L. B. Eyre. New York. [Pp. 5, 15, 42, 101, 104.]

NORRIS, J. (of Bemerton); *see* POWICKE, F. J. (1894).

OWEN, R. (1894). *The Life of Richard Owen.* 2 vols. (Revised by C. D. Sherborn, with a contribution by T. H. Huxley.) London. [P. 107.]

OXENFORD, J.; *see* FISCHER, K. (1857).

PAGEL, W. (1951). William Harvey and the purpose of circulation. *Isis,* vol. XLII, pp. 22–38. [P. 39.]

PALLIS, M.; *see* GUÉNON, R. (1945).

PAPPUS; *see* HULTSCH, F. (1876–8).

PATER, W. (1888, 1st ed. 1873). *The Renaissance.* London. [P. 123.]

PATRICK, G. T. W. (1888). A further study of Heraclitus. *Amer. Journ. Psych.* vol. I, 1887–8, pp. 557–690. [Pp. 12, 106.]

PATRICK, G. T. W. (1889). *The Fragments of the Work of Heraclitus of Ephesus.* (Reprinted from Patrick, G. T. W. (1888).) Baltimore. [Pp. 12, 106.]

PFEIFFER, F. (1861). *Das Buch der Natur von Konrad von Megenberg...* herausgegeben von Dr F. Pfeiffer. Stuttgart. [P. 39.]

PFEIFFER, F. (1924). *Meister Eckhart*. (Leipzig, 1857), trans. by C. de B. Evans (1924, 2 vols.; 2nd impression, title-page date 1947, but not issued until 1949, 1 vol.). London. [Pp. 16, 116.]

PICKEN, L. E. R. (1950). Fine structure and the shape of cells and cell-components (in Discussion on morphology and fine structure). *Proc. Linn. Soc. Lond.* vol. CLXII (1949–50), pp. 72–6. [P. 9.]

PLATO (translations); *see* CORNFORD, F. M. (1935), (1937), (1939); JOWETT, B. (1871).

PLOTINUS; *see* BRÉHIER, E. (1924–38), *and* DODDS, E. R. (1923).

POINCARÉ, H. (1908). *Science et Méthode*. Paris. [Pp. 11, 18.]

POINCARÉ, H.; *see* LEIBNIZ, G. W. (1930).

POIRET, J. L. M.; *see* TURPIN, P. J. F. (1820).

POLLOCK, Sir F. (1899). *Spinoza: his Life and Philosophy*. 2nd ed. London. [Pp. 77, 83.]

POLLOCK, Sir F.; *see* CLIFFORD, W. K. (1879).

POOLE, R. L. (1920). *Illustrations of the History of Medieval Thought and Learning*. 2nd ed. London. [P. 31.]

POWER, H. (1664). *Experimental Philosophy, In Three Books*. London. [P. 13.]

POWICKE, F. J. (1894). *A Dissertation on John Norris of Bemerton*. London. [P. 114.]

PRANTL, C. (1855–70). *Geschichte der Logik im Abendlande*. Leipzig. [P. 94.]

PROCLUS; *see* DODDS, E. R. (1933).

RABIN, C.; *see* SINGER, C., and RABIN, C. (1946).

RENAN, E. (1852). *Averroès et l'Averroïsme*. Paris. [Pp. 10, 113, 114.]

RENAN, E. (1883). *Souvenirs d'enfance et de jeunesse*. II. IV. *Le séminaire d'Issy*. Paris. [P. 52.]

REYNOLDS, Sir J. (1797). *Works*. Ed. by E. Malone. 2 vols. London. [Pp. 14, 19, 91.]

RICHARDS, I. A. (1932). *Mencius on the Mind*. London. [Pp. 68, 118, 121.]

RICHARDS, O. W.; *see* ROBSON, G. C., and RICHARDS, O. W. (1936).

RITCHIE, A. D. (1936). *The Natural History of Mind*. London. [P. 91.]

RIVAUD, A. (1937). Remarques sur le mécanisme cartésien. In the *Recueil* for the Tercentenary of Descartes's *Discours de la Méthode*, published by *La Rev. Phil.*, pp. 290–306. Paris. [P. 116.]

ROBERTS, M. (1920). *Warfare in the Human Body*. London. [P. 105.]

ROBINSON, R. (1941). *Plato's Earlier Dialectic*. Cornell University Press, New York. [P. 28.]

ROBINSON, R.; *see* JAEGER, W. (1948).

ROBSON, G. C., and RICHARDS, O. W. (1936). *The Variation of Animals in Nature*. London. [P. 103.]

ROSS, G. R. T.; *see* HALDANE, E. S., and ROSS, G. R. T. (1911–12).

ROSS, W. D. (Sir D.) (1937). *Aristotle*. 3rd ed. London. [P. 107.]

ROTH, L. (1924). *Spinoza, Descartes and Maimonides*. Oxford. [P. 70.]

RUGGIERO, G. DE; *see* DE RUGGIERO, G.

RUSSELL, E. S. (1916). *Form and Function*. London. [P. 93.]

RUSSELL, E. S. (1930). *The Interpretation of Development and Heredity: a study in Biological Method*. Oxford. [Pp. 101, 102.]

RUSSELL, E. S. (1932). Conation and perception in animal learning. *Biol. Rev.* vol. VII, pp. 149–79. [P. 116.]

RUSSELL, E. S. (1934). *The Behaviour of Animals. An Introduction to its Study.* London. [P. 93.]

RUSSELL, E. S. (1936). Form and function. A historical note. *Folia Biotheoretica,* ser. B, no. 1, 12 pp. Leiden. [P. 93.]

RUSSELL, E. S. (1945). *The Directiveness of Organic Activities.* Cambridge. [P. 101.]

SAINT-HILAIRE, G. É.; *see* GEOFFROY SAINT-HILAIRE, É. (1830).

SCHILLER, F. C. S. (1931). *Formal Logic.* 2nd ed. London. [P. 105.]

SCHMID, G. (1935). Ueber die Herkunft der Ausdrücke Morphologie und Biologie. *Nova Acta Leopoldina,* N.F., Bd. 2, pp. 597–620. [P. 34.]

SCOTT, D. H. (1925). German reminiscences of the early 'eighties. *New Phytol.* vol. XXIV, pp. 9–16. [P. 121.]

SCOTT, G. (1914). *The Architecture of Humanism.* London. [Pp. 11, 50.]

SENECA; *see* CLARKE, J., and GEIKIE, Sir A. (1910).

SHARROCK, R. (1672). *The History of the Propagation and Improvement of Vegetables By the concurrence of Art and Nature.* 2nd ed. Oxford. [P. 52.]

SHELDON, W. H. (1918). *Strife of Systems and Productive Duality.* Harvard Univ. Press, Cambridge, Mass. [P. 96.]

SHERBORN, C. D.; *see* OWEN, R. (1894).

SINGER, C. (1917). The scientific views and visions of Saint Hildegard. *Studies in the History of Science,* vol. I, pp. 1–55. Oxford. [Pp. 38, 39.]

SINGER, C. (1925). *The Evolution of Anatomy.* London. [P. 5.]

SINGER, C. (1931). *A Short History of Biology.* Oxford. (2nd ed. 1950, *A History of Biology.* London.) [Pp. 34, 96.]

SINGER, C. (1941). *A Short History of Science to the Nineteenth Century.* Oxford. [Pp. 31, 39, 46, 84.]

SINGER, C., and RABIN, C. (1946). *A Prelude to Modern Science. Being a Discussion...of the 'Tabulae anatomicae sex' of Vesalius.* Cambridge. [P. 5.]

SINGER, D. W. (1950). *Giordano Bruno: his Life and Thought. With Annotated Translation of his Work on the Infinite Universe and Worlds.* New York. [Pp. 106, 110, 111, 118.]

SMITH, N. KEMP (1923). *A Commentary to Kant's 'Critique of Pure Reason'.* 2nd ed. London. [P. 109.]

SMITH, N. KEMP (1933). *Immanuel Kant's Critique of Pure Reason.* Translation. London. [Pp. 22, 66, 78, 85, 87, 118, 123, 124.]

SMITH, N. KEMP (1947). Ed. *Hume's Dialogues concerning Natural Religion.* 2nd ed. London and Edinburgh. [Pp. 91, 112.]

SMUTS, J. C. (1926). *Holism and Evolution.* London. [P. 53.]

SNYDER, A. D. (1929). *Coleridge on Logic and Learning.* New Haven, U.S.A., and London. [Pp. 117, 119.]

SPAMPANATO, V.; *see* BRUNO, G. (1923–7).

[SPINOZA, B. DE] 'B.D.S.' (1677). *Opera posthuma.* (No place-name.) [Pp. 21, 70, 72, 73, 74, 84, 85, 86, 101.]

SPINOZA, B. DE; *see* VLOTEN, J. VAN, and LAND, J. P. N. (1882–3); WHITE, W. HALE, and STIRLING, A. H. (1899), (1930); WOLF, A. (1910).

STACE, W. T. (1920). *A Critical History of Greek Philosophy.* London. [Pp. 58, 80, 84.]

STACE, W. T. (1924). *The Philosophy of Hegel. A Systematic Exposition.* London. [P. 110.]

STACE, W. T. (1932). *The Theory of Knowledge and Existence.* Oxford. [Pp. 31, 78, 86, 90.]

STEPHEN, L.; *see* CLIFFORD, W. K. (1879).

STIRLING, A. H.; *see* WHITE, W. HALE, and STIRLING, A. H. (1899) *and* (1930).

STOCKS, J. L. (1938). *Time, Cause and Eternity.* London. [P. 114.]

STURT, H.; *see* METZ, R. (1938; reimpression, 1950).

SUZUKI, D. T. (1927). *Essays in Zen Buddhism* (First Series) London. (Published for the Eastern Buddhist Society, Kyoto, Japan.) [P. 55.]

TANNERY, P.; *see* DESCARTES, R. (1897–1910).

TAYLOR, A. E. (1918). The philosophy of Proclus. *Proc. Arist. Soc.*, N.S., vol. XVIII, pp. 600–35. [P. 103.]

THOMPSON, Sir D'ARCY W. (1917). *On Growth and Form.* (2nd ed. 1942). Cambridge. [Pp. 34, 88.]

THOMPSON, W. R. (1937). *Science and Common Sense.* London. [Pp. 58, 81, 125.]

THORBURN, W. M. (1918). The myth of Occam's Razor. *Mind*, vol. XXVII, pp. 345–53. [Pp. 88, 90.]

THOULESS, R. H. (1938). Eye and brain as factors in visual perception. *Pres. Add. Sect. J. (Psychology) Brit. Ass. Adv. Sci. Rep.* Cambridge, pp. 197–212. [P. 115.]

TOYNBEE, P.; *see* DANTE ALIGHIERI (1900).

TREDENNICK, H. (1933). *Aristotle. The Metaphysics*, I–IX. Translation. Loeb Classical Library. London. [P. 37.]

TROLL, W. (1928). *Organisation und Gestalt im Bereich der Blüte* (Monogr. wiss. Bot. 1.) Berlin. [Pp. 43, 122.]

TROLL, W. (1935, etc.). *Vergleichende Morphologie der höheren Pflanzen.* Berlin. [Pp. 25, 34.]

TURPIN, P. J. F. (1820). Iconographie Végétale. In Poiret, J. L. M., *Leçons de Flore*, vol. III, Paris. [P. 121.]

URBAN, W. M. (1939). *Language and Reality.* London. [P. 31.]

VARENDONCK, J. (1921). *The Psychology of Day-dreams.* London. [P. 21.]

VLOTEN, J. VAN, and LAND, J. P. N. (1882–3). *Benedicti de Spinoza Opera.* 2 vols. Hague. [P. 101.]

WALEY, A. D. (1934). *The Way and its Power. A Study of the Tao Tê Ching and its Place in Chinese Thought.* London. [P. 11.]

WALEY, A. D. (1939). *Three Ways of Thought in Ancient China.* London. [P. 37.]

WALLACE, W. (1874). *The Logic of Hegel.* Trans. from the *Encyclopaedia of the Philosophical Sciences*, with Prolegomena. Oxford. [P. 70.]

WALLAS, G. (1926). *The Art of Thought.* London. [P. 21.]

WALSH, W. H. (1946). Hegel and Intellectual Intuition. *Mind*, vol. LV, pp. 49–63. [P. 46.]

WALSH, W. H. (1947). *Reason and Experience.* Oxford. [Pp. 22, 102, 123.]

WARDLAW, C. W. (1950). Organogenesis in ferns: evidence relating to growth centres and physiological fields (in Discussion on morphogenesis). *Proc. Linn. Soc. Lond.* vol. CLXII (1949–50), pp. 13–18. [P. 108.]

WESTAWAY, F. W. (1919). *Scientific Method.* 2nd ed. London. [P. 58.]

WHEELER, R. H. (1935). Organismic *vs.* mechanistic logic. *Psychological Rev.* vol. XLII, pp. 335-53. [P. 101.]

WHITE, W. HALE, and STIRLING, A. H. (1899). *Benedict de Spinoza. Tractatus de Intellectus Emendatione.* Trans. by W. H. W., revised by A. H. S. London. [P. 70.]

WHITE, W. HALE, and STIRLING, A. H. (1930). *Benedict de Spinoza. Ethic.* Trans. by W. H. W., revised by A. H. S. 4th ed. Oxford. [Pp. 72, 73, 84, 85, 86, 94.]

WHITTAKER, Sir E. T. (1946). *Space and Spirit.* London and Edinburgh. [Pp. 29, 32.]

WHITTAKER, Sir E. T.; *see* NEWTON, Sir I. (1931).

WIGHTMAN, W. P. D. (1934). *Science and Monism.* London. [P. 104.]

WIGHTMAN, W. P. D. (1950). *The Growth of Scientific Ideas.* Edinburgh and London. [Pp. 24, 31.]

WILKINSON, E. M. (1949). 'Tasso—ein gesteigerter Werther' in the light of Goethe's principle of 'Steigerung'. An inquiry into critical method. *Mod. Lang. Rev.* vol. XLIV, pp. 305-28. [P. 97.]

WILLIAMS, D. (1947). *The Ground of Induction.* Harvard Univ. Press, Cambridge, Mass. [Pp. 26, 124.]

WILLIS, T. (1664). *Cerebri anatome.* London. [P. 34.]

WISDOM, J. O. (1952). *Foundations of Inference in Natural Science.* London. [Pp. 26, 27.]

WOLF, A. (1910). *Spinoza's Short Treatise on God, Man, and his Well-being.* Trans. and ed. with a Life of Spinoza. London. [Pp. 66, 94.]

WOLF, A. (1922). Spinoza the Conciliator. *Chron. Spinozanum,* vol. II, pp. 3-13. [P. 99.]

WOLFSON, H. A. (1934). *The Philosophy of Spinoza.* 2 vols. Harvard Univ. Press. Cambridge, Mass. [P. 37.]

WOOD, L. (1940). *The Analysis of Knowledge.* London. [Pp. 66, 125.]

WOODGER, J. H. (1929). *Biological Principles. A Critical Study.* London. [Pp. 23, 78, 90, 103.]

WOODGER, J. H.; *see* BERTALANFFY, L. VON (1933).

WORDSWORTH, W. (1800). *Lyrical Ballads.* 2nd ed. vol. I. London. [P. 32.]

WRIGHT, G. H. VON (1951). *A Treatise on Induction and Probability.* London. [Pp. 26, 29, 31, 47.]

YOUNG, J. Z. (1938a). The evolution of the nervous system and of the relationship of organism and environment, pp. 179-203 in *Evolution. Essays on Aspects of Evolutionary Biology,* pres. to E. S. Goodrich on his seventieth birthday. Ed. by G. R. de Beer. Oxford. [Pp. 103, 105.]

YOUNG, J. Z. (1938b). Contribution to discussion on the mechanism of evolution (Brit. Ass. Adv. Sci., Aug. 19, 1938) *Nature,* vol. CXLII, Sept. 17, p. 515. [P. 103.]

YU-LAN, FUNG; *see* FUNG YU-LAN.

ZIMMERMANN, W. (1930). *Die Phylogenie der Pflanzen: ein Überblick über Tatsachen und Probleme.* Jena. [P. 26.]

INDEX

(For those authors' works not included in this index, see page references in
List of Books and Memoirs Cited, pp. 127 *et seq.*)